P9-EJI-089

# THE CLOCK MIRAGE

Also by Joseph Mazur

*Euclid in the Rainforest: Discovering Universal Truth in Logic and Math* (2006)

*Zeno's Paradox: Unraveling the Ancient Mystery behind the Science of Space and Time* (2008)

*What's Luck Got to Do with It? The History, Mathematics, and Psychology of the Gambler's Illusion* (2010)

*Enlightening Symbols: A Short History of Mathematical Notation and Its Hidden Powers* (2014)

*Fluke: The Math and Myth of Coincidence* (2016)

*Number: The Language of Science* (edited, 2007)

# THE CLOCK MIRAGE

## Our Myth of Measured Time

Joseph Mazur

Yale UNIVERSITY PRESS
New Haven & London

Published with assistance from the foundation established in memory of Calvin Chapin of the Class of 1788, Yale College.



Yale University Press books may be purchased in quantity for educational, business, or promotional use. For information, please e-mail sales.press@yale.edu (U.S. office) or sales@yaleup.co.uk (U.K. office).

Epigraph (p. vi): From correspondence with Tasneem Zehra Husain about her forthcoming book *Repaint the Sky,* a collection of "bedtime stories" intended for her son, who would trustingly absorb them now and more deeply appreciate them at a later age. This quotation is an excerpt of the first story and is published with her permission.

Set in Galliard and Copperplate 33 types by Integrated Publishing Solutions. Printed in the United States of America.

Library of Congress Control Number: 2019945867
ISBN 978-0-300-22932-5 (hardcover : alk. paper)

A catalogue record for this book is available from the British Library.

This paper meets the requirements of ANSI/NISO Z39.48-1992 (Permanence of Paper).

10 9 8 7 6 5 4 3 2 1

TO MY FIVE EXTRAORDINARY GRANDCHILDREN,
SOPHIE, YELENA, LENA, NED, AND ATHENA.
EVERY DAY, THEY TEACH ME WHAT TIME IS.

Once, before Time, in a place beyond Space, nothing ever happened. You may think nothing happened because there was no time, and it takes time for things to happen—and you may be right. Then again, perhaps there was no time because nothing changed and there was no way to mark any moment as different to the one before it, or after it, as Time is used to do. So it could just be that there was no time because there was no reason for time. Everyone likes to be wanted.

—Tasneem Zehra Husain

## CONTENTS

### PART III. THE PHYSICS



### PART IV. THE COGNITIVE SENSES



### PART V. LIVING RHYTHMS

## PREFACE

Now, at an age when time is increasingly precious, I wonder why the minute hands of my life seem to move so fast that I sense gaps in place of stolen intervals of my own awareness. So here is a book about time, looking at its many folds, its history, its personal impact, the ways it is perceived, its connections to memory and destiny, its slowness, and its speed. A glimpse at what time really is.

You would expect a book on time to come from the physicist, the philosopher, the chronologist, or the clockmaker. I'm neither a chronologist nor a clockmaker, neither philosopher nor experimental psychologist. I am, however, a mathematician and science journalist embedded with the troupes of real scientists and scholars who productively guide their careers through laboratory experiments examining the depths of knowledge, in particular on the question of time, which—at its root—is partly mathematical, partly conceptual, and significantly imaginary.

*Time* is the most frequently used noun in the English language, yet it is a word that eludes almost every attempt at giving it adequate meaning. If it's a thing, it is a stubborn one. Dictionaries have never been able to peg it. They struggle to give it universal clarity. Their best descriptions give *time* as "a point or period when something occurs," or "the thing that is measured as seconds, minutes, hours, days, years, etc.," or "what clocks measure." My 2,662-page *Merriam-Webster* tries to pin the definition down with a 1,756-word entry. Tries!

It seems that time is a human invention that measures the lines of our existence. We are born with almost no real sense of it. We have clocks, calls to prayer, appointments to keep, places to be, deadlines to meet, expectations, times to eat, and times to sleep. However, since time is neither an event nor an entity that can be perceived, as Immanuel Kant and so many other philosophers have noted, the notion of time itself slips through every attempt to corral it.

Philosophers and natural scientists have been theorizing about time for more than two thousand years. Psychologists have been experimenting with the sense of time for not much more than a hundred years. But it is the public that actually uses and understands time with a reasonable working sense of what it is. Time is often personal, going with the job, the vacation, the triumphs and failures, the solitudes and friendships, self-reflections, illness and fitness, expectations and delays. For the farmer, it is the long spread between planting and harvest. For the racer, it is the split second before the finish. So I gathered the public's intelligent impressions of time from many of the best-informed sources—clockmakers, long-haul truckers, astronauts living on the International Space Station, Olympic racers who won races by less than a hundredth of a second, intercontinental pilots who regularly cross multiple time zones, prisoners who spent months in solitary confinement, jockeys, factory workers in China fitting little screws into iPhones on an assembly line, and hedge fund traders who make suitcases of money by the second. My interviews concentrated on the experience of sensing time's passing while leaving open any questions that had peripheral entanglements with boredom. The results were surprising and diverse in many ways. The public thinks of time differently than physicists and philosophers, who have been justly swayed by the dictates of mathematics. It is as if there are two parallel homonyms that are coincidentally spelled alike, one signifying a public notion, the other a scientific conception. Connected at their roots, they have common threads tied to a clear, unified concept centering on a sketchy idea that time is nothing more than an invented organizing tool enabling civilizations to function with order while empowering science to explore the universe with consistency.

The public impression is that time goes on in the background of life, as a kind of ticking clock independent of one's birth, an imaginary clock continuing forever, marking particular dates along in the memory of the universe. Some folks feel that the man-made clocks that surround us in digital readouts on smartphones tell us absolute time. The predecessor of the iPhone was the twentieth-century mass-produced battery watch, and before that there was the windup watch. With relatively cheap and available watches, time then became the common social regulator. One ate, not when hungry, but rather when the clock called for mealtime; one slept, not when tired, but rather when the clock permitted bedtime.[1] Abstract time became the new medium of existence. Organic functions themselves were regulated by it to make us feel time fly when we are filled with happy and interesting experiences, and linger when our experiences are dull. When lots of things happen within some timespan the memories give an extended telescoped impression of duration. On the other hand, think of those monotonous times we've all spent in waiting rooms of doctors' offices with nothing worthwhile to read. Talk to any one of the more than two thousand prisoners who have been exonerated since 1989 for crimes they did not do. Talk to them about their memories of incarceration, about their illusions of time's passing. Then talk to the prison guards and wardens about their sense of time. They get to go home at the ends of their workdays. They too spend their days in those prisons, yet they have different notions of time.

Such diverse public impressions make the word elusive. Time is the most ubiquitous topic that has ever entered the literature of the world. Its meaning is endless. In one peculiar sense, we are all writing about time. No wealth of books can possibly exhaust the subject. We can't get enough of it. This book follows the more than two hundred thousand published essays and books that have expressed views on the meaning of time since Aristotle's *Physics*. It will not be the last.

*The Clock Mirage* is not specifically about time travel, the age of the universe, parallel universes, the fourth dimension, any particular time topics of physics, or how to fix a clock. Rather, it is a book that tries to answer the question of what time is by the broadest meaning. But time is a tricky concept, built from a combination of concrete examples coming from culture, education, environmental experiences, and innate senses of its passing. Where will we find a thread of time's passing without a clock to help? No wonder the concept is so ungraspable. There is nothing quite like it, beyond the iconic representational, vaporous picture of some kind of clock in the mind. Perhaps the Mad Hatter in *Alice's Adventures in Wonderland* had it right when he personified time, believing it to be a "him" not an "it": "If you knew Time as well as I do," said the Hatter, "you wouldn't talk about wasting it. It's him." Why not a him? In Hesiod's *Theogony,* the ancient Greek mythic cosmology explaining the creation of the world, time was deified as Chronos (or Cronus). And haven't we all heard of Father Time, the old bearded man who carries a scythe and hourglass?

These chapters will take you on a curious tour of time through the eyes and mind of a person who looks at the world as being simple while knowing that it is actually more complex than can be quickly understood. It covers the story of time, whether "he," "she," or "it," to tackle an answer to the question of what time really is. Is there a secret, a theory, or a nucleus of knowledge that can bring time into focus, or at least into some scope for sharp inspection? Perhaps time is, as Julian Barnes suggests in his novel *The Sense of an Ending,* just a relationship with memory: "I know this much: that there is objective time, but also subjective time, the kind you wear on the inside of your wrist, next to where the pulse lies. And this personal time, which is the true time, is measured in your relationship to memory." Is it that, just memory, and nothing more? Or is it an illusion of invisible posts between the past, present, and future, connecting memories to expectations, to give the mind anticipation, hope, ambition, and, perhaps, a belief in destiny? Years ago, I put these questions to the British mathematician and science historian Gerald James Whitrow, who wrote several books about time, including *What Is Time?* and *The Nature of Time.* He told me, as if conveying a distinguished proverb, that if one wants to know what time is, one must first know what a clock is. When he saw that I didn't understand, he stressed, "I mean what a clock *truly* is." Taking his advice, I learned as much as I could about clocks and, by that learning, soon came to realize that time has something to do with our pulses, yes; but it has more to do with our cells and brains, things that update memories to tell our bodies that we are in the rhythms and beats of being alive.

Welcome to *The Clock Mirage.*

# PART I

# THE MEASURES

# 1

## TRICKLING WATERS, SHIFTING SHADOWS (TELLING TIME)

> The gods confound the man who first found out
> How to distinguish hours! Confound him too,
> Who in this place set up a sun-dial,
> To cut and hack my days so wretchedly
> Into small portions—
> —*Plautus,* The Woman Twice Debauch'd

Whatever it is now, time's measure must have once been roughly calibrated by the mortal human lifespan. The baby grows, learns to walk, becomes a child, learns to survive, grows strong, becomes an adult, grows old, and ultimately dies. All other cycles—the motions of the moon, sun, and earth—are simply parallel measures of the rounds of human existence. The year, one complete cycle of the seasons, is the simplest marking of the human path through its own presence. A day in the year is easy to measure as one sunup to the next but is hard to notice, like a day's change in the growth of a child. Yet the child first gets to know time as repeating days. Our day is a convenient accident. Thank the gravitating clouds of molecular dust that collapsed by accretion 4.6 billion years ago to become a planet that spins on its axis as it orbits the sun to give us the hours of the day and the number of days in one year. But spin and orbit are not coordinated quite well enough for the number of days of the year to be a simple whole. So, dividing the day is not so simple as it might appear.

The first idea for dividing the day into twenty-four hours, as well as the use of the word *hour* as a partial interval of a day, occurred during the reign of Nebuchadnezzar II (ca. 605–562 BC), the Chaldean king of the Neo-Babylonian Empire responsible for the destruction of the Jerusalem Temple.[1] The Greek historian Herodotus used the word *ορα* (hour) in his *Histories*, written in about 440 BC. We know from *The Histories* that the Greeks learned about the sundial and the division of days into twelve parts from the Babylonians.[2]

Our twenty-first-century clocks don't just tell us what time it is but signal many of our daily appointments with vibrations—*time for the next pill, time to stand up and to exercise, time to leave for that 1:00 p.m. meeting crosstown, 'cause there's a delay on the A line, time to* . . . Yet it seems that, even in those sundial days of ancient Rome without a breakdown to minutes, our days were already governed by time markers. The epigraph above comes from a fourteenth-century book called the *Comedies of Plautus.* It continues:

> —When I was a boy,
> My belly was my sun-dial: one more sure,
> Truer, and more exact than any of them.
> This dial told me when 'twas proper time
> To go to dinner, when I aught to eat—
> But, now-a-days, why, even when I have,
> I can't fall to, unless the sun give leave.
> The town's so full of these confounded dials,
> The greatest part of its inhabitants,
> Shrunk up with hunger, creep along the streets.[3]

Plautus was a second-century BC playwright of comedies, the Molière of his day. A footnote in the English translation of *Comedies* tells us that Plautus lived in the time of the Second Punic War (218–201 BC), the war under Hannibal's leadership between Carthage and the Roman Republic, and that the first sundial in Rome was installed a few decades earlier, in 254 BC. Some ancient sources tell us that at that time there was just one sundial installed, that it was brought to Rome from Sicily, and that Plautus exaggerated the number of sundials in Rome at a time when he was in ill humor.[4] However, we do know of thirteen ancient sundials that were unearthed in Rome in the last century.[5] In the Middle Ages there must have been thousands of them between England and Greece, since the sundial was one of the simplest ways of telling time with an accuracy as good as needed. Sundials were everywhere—on public buildings, courtyards, and public squares.[6] We know of just a few because larger sundials were generally made from limestone that weathered away over time.

For many centuries, the sundial and the water clock (a tank filled with water leaking at a nearly constant rate with indicators mechanically controlled by floating bobs connected to levers marking time) governed the passing days as fuzzy passages of time from dawn to dusk with hardly any care of precision. The technical difficulty of using the sun to break time down into more precise intervals, such as hours, was then, as always, linked to the complex vernal shifts between the seasons.

Shadows from trees and rocks must have played with human curiosity since the beginning of human existence, especially when they moved as days continued toward nights. Human curiosity would have given someone the idea of watching the shadow of a stick in the ground. A few pebbles on the ground would have marked the position of the stick's shadow. Such a remarkable revelation must have taken place far earlier in the human timeline than 2000 BC. But if and when it did happen, it would have remained mostly just a momentary curiosity, not a general community event that became part of local civilian timekeeping.

With improvising cleverness, the mathematics of lengths and positions of stick-in-the-ground shadows could be calibrated to assign numbers that mark moments of a day's passing. Find solar noon and start from there. Next, get a more reliable stick, perhaps an obelisk, perhaps a pyramid. That's all a primitive sundial needs. It's not surprising, therefore, that the idea of a sundial as a time gauge that was commonly used well into the Middle Ages goes back at least four millennia to Babylonia, ancient India, China, and Egypt. Its utility is one of the prophecies of Isaiah (38:8): "Behold, I will cause the shadow on the steps, which is gone down on the dial of Ahaz with the sun, to return backward ten steps." Since Ahaz is mentioned, Isaiah must be talking about a time near Ahaz's reign as king of Judah (732–716 BC). So, it is arguably evident that sundials were known for centuries from Babylonia to Greece.

The Egyptians, as far back as 2150 BC, divided night into segments, yet there is no evidence of a daylight division until about seven hundred years later.[7] A twelfth-century BC Egyptian papyrus known as the Cairo Calendar listing lucky and unlucky days of the year ("Someone born on this day will die by crocodile" and so on) also lists the number of hours of daylight and darkness each day of the year, always totaling twenty-four.[8] It seems to be a remarkable coincidence that the circumference of the world at the equator is very close to 24,000 miles, actually 24,873.6 miles by NASA's estimate. It's a total accident, a coincidence of measurement coming from the fact that the mile turns out to be so close to 1/24000 of the girth of the earth at the equator, a measurement that early Egyptians did not know about, even though they did divide the time of day and night by twenty-four. A coincidence of numbers, given that any resemblance to the modern mile was not known before ancient Roman armies marked their marching journeys in thousand-pace markers. It's as if the universe has given us a gift of convenience.

In Roman times the mile was called *mille passus* (a thousand paces) or, more standardized, the total distance of one thousand steps of the left foot in a Roman soldier's march, which averaged to 5,000 feet (a foot being the average length of a Roman soldier's foot) and was much later adjusted to be standardized at 4,860 English feet. Distance units generally have their origins in either sizes or motions of the human body. In ancient Egypt the *djeser,* or the bent arm, was defined as either four palms or sixteen digits.[9] Over time, the mile, like the foot, varied from country to country and town to town. The modern mile came into use after the fall of the Roman Empire through the English furlong, literally the length of field an ox could plow without resting, a unit that was already standardized at 660 feet. In 1593, in the thirty-fifth year of Elizabeth I's reign, Parliament decreed, "A Mile shall contain eight Furlongs, every Furlong forty Poles, and every Pole shall contain sixteen Foot and an half."[10] Thus, under that statute, the English "statute mile" was declared to be eight furlongs, or 5,280 feet. We could have proclaimed the mile to be 1/24000 of the circumference of the earth at the equator at the start, had we known more about the earth's girth, and still have the mile come out to be relatively close to 5,280 feet.

The convenient division of the circumference of the earth by 1,000 in miles gives us twenty-four division points to mark time as the world turns through a day. Of course, it's not really a coincidence. Had that girth in miles been any number other than 24,000, we could have divided by some number to get twenty-four division points, that handy number having many integral divisions. We owe our twenty-four-hour split of the day to the Babylonians and Egyptians, when the clock was either a sundial or a tank filled with water leaking at a constant rate. Most of the world keeps time in twenty-four divisions while North America, Australia, and the United Kingdom repeat 12 twice a day.

Ancient Egyptians divided time into ten parts per day, with two short parts reserved for daybreak and twilight. Moving shadows on obelisks partitioned days into two parts broken at noon (a division instant that is a precursor to our a.m. and p.m., *ante* and *post meridiem,* division) the one time that could be marked with some precision. Visible stars in the night sky marked and divided nighttime hours symmetrically. Splitting day and night into twelve parts probably came from the twelve moon cycles and the Babylonian division of the night sky into twelve sections. Babylonians certainly recognized the convenience of twelve having many integral factors.

By the time of the death of Alexander the Great in 323 BC, all Mediterranean astronomy followed the Babylonian and Egyptian mathematical tradition of dividing hours sexagesimally (into sixtieths) referring to a small group of thirty-six constellations, each one coming over the horizon at a uniform time. Like 60, the number 36 also has a relatively large number of divisors, and the number 12 happens to be one of them.

Early Egyptians had their sundials and water clocks meter divisions of days. Though there was no concept of "five minutes," there surely was a notion of time passing as the sun continued its journey across the sky from horizon to horizon. There was no so-called afternoon or morning as a division of a day. A day was a continuous passing of light. But as civilization got more and more complex, so did the divisions of the day.

A record of the earliest Egyptian sundials is told in the stories of the battles of the fifteenth-century BC pharaoh Thutmose III, whom modern historians have called the Napoleon of Egypt because of his military genius and expansionist wars. A 2013 archaeological expedition in the Valley of the Kings, where Thutmose was buried, conducted by the Egyptologists Susan Bickel and Elina Paulin-Grothe and their team at the University of Basel, uncovered a limestone disk that is now considered to be the oldest existing fragmented sundial of its kind.[11]

Next came the water clock, the clepsydra. One of particular curiosity dates back to about 145 BC Alexandria.[12] In a stretched sense, this was a hybrid mechanical clock with toothed wheels and a weighted regulator. In a crude way the regulator worked alongside a mechanism that we now call an escapement, a device that alternately catches and releases a stream of motion according to a gauged amount of flow. The modern escapement is also such a disciplined measurer; it counts a flow by the number of catches and releases, but it also transmits an impulse with each catch to give a kick of energy to the system.

In ancient Rome the word for clock, *horologium,* a compound joining of *hora* (hour), and *lego* (to point out), was used for any time-measuring device: a reference to *horologia* could be to a sundial, a clepsydra, or a mechanical clock. In his *Ten Books on Architecture,* Vitruvius, who lived in the first century BC, describes the mechanism of a clepsydra, possibly one invented by Ctesibius of Alexandria in the third century BC.[13] John Farey Jr., an English mechanical engineer, drew Vitruvius's clepsydra in 1819, as illustrated on page 9. As water enters the central tube, C, it raises a float, D, connected to a figure that points to the hour. The significance of this clock is that even though gears in the form of toothed wheels had been used long before the first millennium, the clock that Vitruvius describes is an ingeniously created timepiece that adjusts itself for the seasons by virtue of a spillover that powers a series of gears.[14] It is hard for me to believe that a clock of this sophistication was built before the first millennium; however, Vitruvius says it is so.

In China there was Su Song's clepsydra. In 1088, Su Song, a wizard of science and technology, built a Rube Goldberg contraption several stories high that used a turning waterwheel of buckets that scooped and spilled water to tilt levers that alternately caught and released sprockets attached to the wheel. Su's scoop-and-tilt mechanism counted time by the bucket-load and relayed the count to a gear mechanism in the clock that rotated a celestial globe and jacks that hammered away bells to sound the time.[15] The clock is now gone, but was described in a book written in 1092 titled *Hsin i-hsiang fa-yao,* which literally translates to "Essentials of a New Method for (Mechanizing the Rotation of) an (Armillary) Sphere and a (Celestial) Globe." The entire Su Song clock tower was likely an elabo-

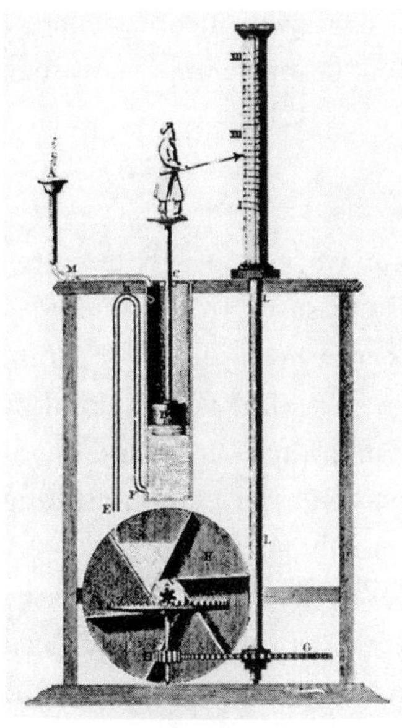

Abraham Rees, "Clepsydra," pl. 3, in *Cyclopædia: or, a Universal Dictionary of Arts, Sciences, and Literature,* vol. 2 (1819): 359

rate astronomical clock controlled by a scooping and spilling bucket contraption acting as a mechanical escapement. There were many wheels with interconnected lugs ringing bells and beating drums. All parts were connected by rods that would make colorful figures pop in and out of doorways while raising and lowering arms with hammers to strike gongs.[16]

Time in China, then, was divided into a hundred parts of the day. In Su's clock, at every fourth part, by a sophisticated mechanism of meshing gears and chains, mannequins holding tablets would exit doors and announce the time of day by ringing bells and beating drums. A vertical shaft turning a globe was gear-meshed to a horizontal shaft connected to a waterwheel that somehow moved in increments of a hundredth part of a day.[17] It is a stretch to suggest that Su's clock truly used the principle of an escapement, yet it did use a scoop and tilt (a somewhat back-and-forth) mechanism to regulate time.

By the seventh century there were several Chinese water clocks displaying time by raising placards, ringing bells, striking gongs, beating drums, and making mannequins appear, disappear, and appear again in different clothes. Word of these ingenious mechanical engineering marvels traveled along the Silk Road to Persia and thence to medieval Europe, where a clock's display and fanciful embellishments were far more important than its accuracy.

---

Throughout history humans have tried many ways to loosely tell time. Sidereal (from the Latin *sidereus,* star) time is measured from the earth's position relative to a fixed star. Solar time is measured by the apparent position of the sun relative to the earth. Mean solar time is computed by finding the time it takes for the sun to pass from noon to noon and dividing that interval by twenty-four. Standard time is an agreed-upon dividing of the earth into twenty-four zones running mostly north-south.

Sidereal time can be measured directly. From earth the sun appears to be in particular positions against the backdrop of stars in the sky as it slowly and nonuniformly moves eastward, taking a year to return to its position against the stars. The apparent path is an imagined great circle called the *ecliptic,* because eclipses happen only when the moon crosses that path. Imagine a plane, the *celestial equator,* cutting through the earth's equator extended through the solar system. There are two points at which the ecliptic crosses the celestial equator. One, which happens around March 20, is the *vernal equinox* (from the Latin *vernalis,* spring). The other, which happens around September 22, is the autumnal equinox. From the earth, those two points in the sky can be pinpointed against the backdrop of the stars.

Now precisely, we can talk about when a day begins. You are reading this book at a particular location. Stand up. Turn to face north. Extend your arms to point east and west. Now imagine a plane through your body and the center of the earth, the *meridian plane.* From observation on earth the vernal equinox will appear to be moving across the sky, because it is the backdrop of the stars that appear to be moving. That vernal equinox will cross the meridian plane every day just once. Each time it crosses, a new day begins. Of course, what we see is the sun rising and falling as the day continues. But we know that the appearance of the sun's rising and falling is just a relative impression that follows by observing the sky from a particular position on our spinning earth. That crossing can be detected with amazing precision. Now that we have a beginning (and

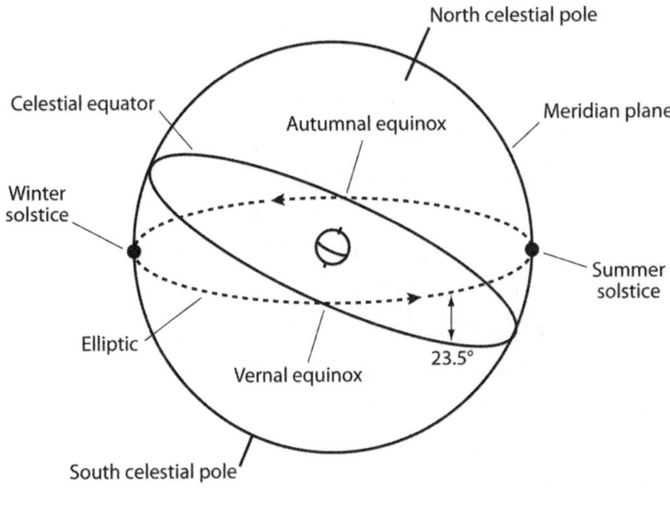


Meridian plane

hence an end) to the day, we can divide that interval evenly by twenty-four to get the hour, those hours by sixty to get the minutes, and those minutes by sixty to get the seconds. Or, we could just as well divide the day in any way we wish to get precise times.

Roman hours of the first century depended on the season. The day was divided into quarters—early morning and afternoon, afternoon, and evening.[18] Winter daytime, when the sun shone for a bit less than nine hours (by our meaning of *hour*), was divided into twelve forty-five-minute segments (by our definition of *minute*). In the summer, the scheme would reverse—a clever scheme that gave one the impression that shorter days in winter had the same number of hours as summer. Clever, though inefficient. They were still using water clocks that had to be reset and calibrated every day by an official timekeeper who took a month to make the complete round of Rome's many water clocks. Even today, in modern New York City, Marvin Schneider, the official clock master, sets and adjusts clocks for all city-owned tower clocks visible from the street.

Countries away from the equator had to account for the seasonal changes in daylight hours. The famous Prague Clock (illustrated in chapter 2), for instance, was originally constructed to mark an eight-hour day, a remnant of times when a town clock meant one of three things: time to wake up for work, time to end the workday, and time to attend church. As civilization moved forward to market economies and to travels that had to be coordinated, time had to become synchronized beyond the local times of village clocks. The pace of life was tough, yet simple, before the machine age. People were mostly agrarian and rural. Labor was largely farming and craftwork that used either hand tools or simple machines driven by whatever animal power a person could afford.

Then came the Industrial Revolution, which completely changed the way time is measured, understood, and felt. The steam engine with its mass-production potentials brought with it the factory, a new way to make and sell goods, a "creature of the modern world, a world it helped create."[19] The noun itself, stemming from the Latin *facere* (to make) referred to any building that housed a mechanized contraption that could make something useful for domestic consumption. The idea originated in Derbyshire, England, in the early eighteenth century and from there spread slowly for the next few decades, when small engines began being enlarged to mammoth-size machines working in tandem to turn out textiles at miles per hour rather than yards per hour.[20] Hamlets and villages of established self-sufficiency lost their strong and young to the growing textile factories that lured workers with promises of good and steady wages. By the next century, the north of England had smoke-blowing, soot-creating factories with thousands of workers working six days a week from dawn to dusk for pay that couldn't keep up with inflation. Engineering advances and increased consumer buying power brought faster machines and longer working hours to speed up production. With hard and mostly grim and boring work, workers felt a sense of relief as the days approached Sunday or a rare vacation day. People who worked hard and long days in the growing industrial regions of Europe were either offspring of farm laborers or farmers themselves, who knew only of the constant needs of the farm, milking and haying, which had no days off. For industrial workers, *weekend* became a word. And as the mechanized age developed, so too did transportation infrastructure, vis-à-vis canals, roads, and rails, along with the express need to know time more precisely than ever before. The clock with just an hour hand was no longer good enough; even the village clock bell that struck on the quarter hour was no longer good enough.

---

Our English word *clock* originated from the German *glocke*, which means bell. By the sixth century, European monasteries had installed mechanical contraptions driven by weights that would awaken a bell ringer, who would in turn ring large tower bells to announce canonical hours—sunrise matins, noon none (the ninth hour after sunrise), evening compline, and nighttime vespers. Numbers on a dial would have been meaningless in a simple civilization that cared only for a few breaks in a day's routine. Breaking the time of day into minutes and seconds came much later, when punctuality in commerce became critical for organized appointments, shipping, and local travel. Such thin divisions required more accuracy than the sundial or water clock could give. A mechanical device was needed, one that could count equally spaced moments without the help of intermittent sunlight or continuously flowing materials such as water, sand, or burning oil. The earliest clocks needed perpetual maintenance for continuity and accuracy. It's not easy to devise a mechanism that is continuous and regular. The human pulse, which Galileo is reputed to have used as a measure of regularity, changes in short intervals. At first thought, it may seem easy to keep a gear moving at a steady, continuous rate. On further thought, it becomes clear that this may be one of the world's most difficult problems. To measure time, there must be something to count.

There was nothing more finely scaled than divisions of days into twenty-four, or twelve, or eight parts, surely nothing close to a division of minutes, or even quarter hours. Even by Galileo's time, punctuality was not important. Dinner invitations would have been as vague as, say, "Come to dinner at sunset." Government meetings with visiting dignitaries would happen whenever the dignitaries would arrive. Clocks worked by dripping water, burning oil, pouring sand, or sunlight. Accuracy was not a word to be used for time measurement. But real time always requires a counting of something. No matter how finely calibrated our clocks are, they are always measuring something discrete—an interval, a repeating signal, a duration between events, the weight or volume of water or sand accumulated in a vessel, or the angle of a stick's shadow from sunrays. This is the heart of the problem: we measure time as a number and think of motion or change as being continuous. For the past nine hundred years almost all mechanical clocks that we know of have used some mechanical principle of counting discrete movements. Some involved gears with mechanisms that counted rotations. But think about how difficult it is to make a gear count, and think about how clever it is to have an arrow turn to a marked number. The clockface with clock hands pointing to marked numbers is still with us, although recently competing with digital displays. The early clockface clocks had only one hand, pointing to the hour. The minute hand came along at the end of the seventeenth century, and soon afterward a third hand, called the second minute hand, appeared, to mark what has ever since been called the *seconds*.[21] That second hand, when first installed, was as superfluous as it is today, but it became basic in the design of eighteenth-century mantelpiece clocks because . . . well, why not have a new feature that is easily built by throwing in a couple more gears? The great advantage of the analog clockface over the digital display is that the clockface gives a sense of the nearby past and upcoming future. Checking the time on a round clockface tells us without computing how much time we have before we need to be someplace else, and it tells us how long it has been since we were doing something else. We get a real sense of past, present, and future just by a glance of the face, a significant utility totally lost in digital displays that read out the almost useless detail of seconds.

Atomic clocks now measure time intervals as small as one trillionth of a second. Interesting to think that, though we have words for times smaller than a trillionth of a second (*femtosecond* = $10^{-15}$ of a second, *zeptosecond* = $10^{-21}$ of a second, *yoctosecond* = $10^{-24}$ of a second), we have no way of accurately measuring such microtimes. Like the human pulse that sends a continuous flow of blood through arteries in distinct, discrete pushes coming from contractions of heart muscle that can be counted, time is both continuous and discrete. Any clock will measure time by counting something discrete in the continuous flow. We have meters and gauges that measure the continuous flow of all sorts of both liquid and gas products. When you fill your car with gasoline the gas flows continuously, and yet the number of tenths of a gallon changes discretely. That measurement comes from a flow meter, a simple rotor in a chamber that the gas flows through before reaching the hose. That rotor "counts" its revolutions, and—just like a clock measuring time—ticks off the amount of gas flowing through in units of tenths of a gallon.

In this century and the last, we have had the need for incredible fine-tuning of time to coordinate stock trades, train commerce, power grids, airline navigation, medical technology, and, let's not forget, sports timing. We have a fairly precise digital readout of time on our cellphones whenever we wish. We were even able to determine the winner of a four-hundred-meter track and field championship in Rome. The year was 1987. There was a clock that could detect the winner, Edwin Moses, who won the race by just two hundredths of a second. And let's not forget the 1998 Winter Olympic Games in Nagano, Japan, when U.S. skier Picabo Street won a gold medal in a women's super-G race by coming in one hundredth of a second faster than Michaela Dorfmeister of Austria!

## INTERLUDE: OLYMPIC RACER WINS BY ONE HUNDREDTH OF A SECOND

Talk to the American alpine skier Picabo Street about her one-hundredth-of-a-second win over Michaela Dorfmeister from Austria in the super-G 1998 Winter Olympic Games. It is not often that the difference between gold and silver is one hundredth of a second. The race was almost as close for the silver: Alexandra Meissnitzer from Austria missed silver by just seven hundredths of a second. Try to imagine the duration of one hundredth of a second. It is shorter than the flap of a hummingbird's wing. Seven hundredths of a second is far shorter than the blink of a human eyelid.

Alpine slaloms are not horse races where environmental conditions are extremely close to even and where jockeys pace each other. Each skier in a slalom has a separate run under variable conditions. A slight headwind gust can slow the timing by more than just seven hundredths of a second. Declaring winners by hundredths of a second ignores the unmerited factor of arbitrary luck. To be fair, all three racers in that 1998 super-G Olympics race should have shared gold. In the last two centuries of horse racing there are many examples of dead heats, when the finishes of two or more competitors were too close to distinguish and the prizes were divided.

At the 2012 Summer Olympic Games triathlon event in London, the Swiss triathlete Nicola Spirig won gold over the Swedish triathlete Lisa Nordén. Their timing was precisely the same, 1:59.48. A photograph showing Spirig ahead by a hair determined the medals, but both Spirig and Nordén ran through the same environmental conditions, although one could say that their footsteps were on different soils. Again, the luck of the step.

On February 17, 2018, at the Daytona International Speedway, Tyler Reddick beat Elliott Sadler in a 357.5-mile-long NASCAR race by 0.0004 seconds—that's four ten-thousandths of a second![1] Such races can be close in part because contestants pace each other. The super-G 1998 Winter Olympics race was a sequence of single racers, each pushing to race at a personal best, without another racer to pace, and yet the split-second endings happened. So what, then, is the meaning of time?

# 2

## RINGING BELLS, BEATING DRUMS (USE OF TIME)

> I want to build a clock that ticks once a year. The century hand advances once every 100 years, and the cuckoo comes out on the millennium. I want the cuckoo to come out every millennium for the next 10,000 years.
> —*Danny Hillis, inventor of the 10,000 Year Clock*

By the end of the twentieth century clocks were everywhere. If you walked along any busy city street, you were never far from scanning glances of clocks. Subway clocks in New York City were centrally calibrated, as were many street clocks of Paris and London. Some streets had large clocks at sidewalk edges, and you were always within range of some store with a clock hanging on a wall that could easily be seen through a front window. Today there are far fewer public clocks, but we are more than ever very closely connected to the most precise time of day by way of phones and computers set to internet provider clocks. We work and play by the clock more than ever, and we are never more than a blink away from a clock display. The clock has become the big brother of time. There are no more excuses.

Time, for the Europeans of the Middle Ages, was very different. In the fourteenth century, when so many wars between kingdoms and duchies were fought, the best clocks were being stolen from one town and brought to another. One clock, famous for its intricate mechanism, was a "miracle of art," a clock that "had not its equal in all Europe . . . a clock which showed the course of the sun, moon, and stars, and the rise and fall of the tides."[1] Richard of Wallingford, abbot of Saint Albans in England, probably built it for the Flemish city of Kortrijk in 1326. In 1382, when the French army entered Kortrijk, Philip II, duke of Burgundy, ordered the splendid clock to be dismantled and brought to Dijon, his capital, where it still is in the tower of the church of Notre-Dame with a clockface and dial mechanism added.[2]

The Dijon miracle-of-art clock might have been the first mechanical clock driven by weights. Another candidate for being first is a truly mechanical clock built for Charles V of France in 1360 by the clockmaker Henri de Vick. Similar to the hybrid Alexandrian clepsydra, it is certainly one of the simplest mechanical clocks, with just an hour hand that makes twelve moves twice around in a twenty-four-hour day and does nothing more than indicate the hours.[3] No bells. No figures coming out of pagodas. No complicated movements. De Vick took eight years to complete the simple clock, very likely forging all the parts himself. After many renovations and repairs, the clock is now in the Palais de Justice in Paris. Weight driven with an escapement, it is the first clock having spaces between the hours divided into five equal parts (suggestively indicating minutes), just like our modern clocks.

These early hybrid clocks were not meant to tell the time with any precision better than giving an approximate hour. In many ways, if we ignore the difficulties of life in those days—wars, famine, diseases, relatively short lifespans—civilization was far more relaxed about time. Being where one was supposed to be within the imprecise hour was good enough. The original clock at Dijon had no dial (a word that came from the Latin *dies,* meaning day). It had hammers that were struck by automata figures, what English-speaking people called "jacks," from the French word *jacquemart.* Jacques Marck was a clock- and lockmaker from the town of Lille who happened to be the grandson of the clockmaker of the Dijon clock. But there are other reasons why we call these hammering figures "jacquemarts." The word could also have come from the Latin *jaccomarchiardus,* which literally means clothing for war, as in a coat of mail. There was a custom in the Middle Ages to station sentries on towers wearing "jacques" to warn of approaching enemies.[4] Whatever the origin, the first hybrid mechanical clocks had jacks that hammered out the hours. If you were in Dijon in the first few years after 1360, you would hear a gong on the hour.

The magnificent Prague Orloj (Prague Astronomical Clock) is a contender for being the oldest working clock in the world. For many years its mechanism was still. Of course, the clockwork itself had been renovated many times and is now modern. But its magnificence in engineering and science is astounding, given that it was first installed almost exactly six hundred years ago.

There are two dials, one a perpetual calendar, the other showing motions of the sun and moon. Just before striking the hour, two windows open to reveal twelve apostles marching inside the windows. A skeletal figure, supposedly Death, stands at the right side of the bigger dial, rings a bell, and turns over an hourglass, while a man standing nearby nods his head. Two other figures are on the left side; one carries a purse, the other, perhaps Vanity, carries a mirror. Every hour on the hour, a cock appears above the big dial and crows three times.

The clock has several curious legends associated with it, at least one of which is completely untrue yet misrepresented in the history books. One involves a master clockmaker, Jan Růže (pronounced *Yan Rouge-eh*), who was enlisted by the town councillors of Prague to repair the clock and to make it a beautiful and unique instrument for the public square. Almost ninety years after the clock was first installed, Růže repaired it, added a calendar dial, and embellished it to include moving jacquemarts. As the story goes, Růže restored and enhanced the clock as perfect machine, so perfect that the councilors worried that he would become so famous that other towns would solicit him to build similar clocks. Concerned that the clock would no longer be unique to Prague, they blinded Růže with an iron pipe. Blind, he asked one of his students to help him sabotage the workings to stop the clock for a whole century in revenge for the councillors' horrible deed.[5]

Another legend relates to Death, the skeleton figure. If the Orloj ever stops running for a long time (whatever that means), the Czech nation will go through bad times, indicated by the skeleton nodding its head. Hope for the nation's survival (as well as the clock's) rests on a boy being born on New Year's night. When the clock starts moving again, that boy, now grown to be a man, must run out of the nearby church when the clock makes its first bell strike and run across the town square into the town hall before the clock makes its last strike. If he makes it, he will have dissolved the skeleton's power, and all will be well again.[6]

It was an era rich in popular dark myths, just as the present era is abundant in conspiracy theories. The truthful history about the Orloj is hard to untangle, since the records are so obscure and overlapping with contradictory notes. Most historians believe that at least one part of the clock was built in 1410 during the reign of Emperor Wenceslaus IV and other parts were assembled in the years that followed, with the full clock being operational in 1490, when Růže remodeled it to become the astronomical clock with all its marvels. Still active, it is a marvel of engineering for any clock of the fifteenth century. It's likely that it was designed and partially built by the imperial clockmaker Mikuláš of Kadaň, but surely the full clock was not built by him, as some history books and articles claim, since Mikuláš died in 1419.

One of the earliest writings containing detailed accounts of medieval horology is *Le Songe du Vieil Pèlerin* (The dream of the old pilgrim), by Philippe de Mézières, from 1389.[7] Written by a French crusading soldier and statesman, it is largely a personal account of European and Near East observed history. In it we find a narrative of a mid-fourteenth-century clockmaker in Padua, one Maistre Jean des Orloges, who built an instrument in brass and copper that, against a backdrop of the zodiac, mimicked the then-supposed circular motions of the planets along with the sun's orbit around the earth.

> On any given night, we see clearly in what sign and degree are the planets and the stars of the heavens; and this sphere is so cunningly made, that notwithstanding the multitude of wheels, which cannot well be numbered without taking the machinery to pieces, their entire mechanism is governed by one single counterpoise, so marvellous that the grave astronomers from distant regions come with great reverence to visit the said Maistre Jean and the work of his hands; and all the great clerks of astronomy, of philosophy, and of medicine, declare that there is no recollection of a man, either in written document or otherwise, who in this world has made so ingenious or so important an instrument of the heavenly movements as the said clock.[8]

Prague clock. Photo: Carol Ann Lobo Johnson, 2014

It was probably not the first mechanically driven astronomical clock. In Dante's *Paradiso,* canto 24, competed in 1320, we find a description of what must be a gear-driven clock.

> And as the wheels in works of horologes
> Revolve so that the first to the beholder
> Motionless seems, and the last one to fly.[9]

It seems that within Dante's lifetime there were gear-driven clocks, although they could have been clepsydrae, not purely mechanical. The word *clock* in those days did not distinguish differences between sundials, clepsydrae, and mechanical clocks. The clepsydrae of Greek and Roman times could have been mechanical, and probably were.

And so, it is very difficult to know when the first truly mechanical clock was invented. By mechanical clock here I mean a clock driven not by falling water but by falling weights, expanding springs, or pendulums. It is true that some water-driven clocks had mechanical features such as slowly rotating wheels with cords attached to weights that were lowered by floats in water vessels, but such clocks were not counting events. They were not accumulating numbers but rather marking semicontinuous motions. Some scholars credit the first mechanical clock to the Chinese mathematician and monk Yi Xing, who did invent a clock with a crude escapement mechanism in 715 that indicated not just the hours of a day but also positions of the stars by way of a turning celestial globe. But it's clear that Yi Xing's clock was a clepsydra with mechanical linkages that could turn the globe. In eighth-century north central China (west of modern Shanghai) there were all sorts of mechanical toys driven by water, many like the ones that are sold now as beach and tub toys for children, but none known that were controlled by escapements that could calibrate the hours of a day.[10] When thirteenth-century Europe needed more dependable and finer divisions of the day, European craftsmen took up the challenge and developed mechanical clocks with elaborate, ingenious mechanisms. In particular, a perfected escapement was designed to serve the function of breaking continuous motion into short, equal intervals that could be counted. An escapement *is* the mechanical clock. All other parts are simply indicators of time, time reporters with instructions from the escapement.

Imagine a wheel attached to an axle. Now imagine that there is a rope wrapped around the axle with a weight attached to one end. If the wheel were allowed to rotate freely, the rope would unwind to let the weight fall downward by the forces of gravity hampered by friction in the bearings of the axle. Ideally, we would want a gear with sixty teeth to move around at a rate of one tooth per minute, taking one hour to make the complete turn. There is no trick to this, considering that a smaller gear could be made to mesh with the larger and turn at a rate that could be adjusted by the ratio of diameters of the two gears. But it is the speed of the escapement mechanism that will determine whether the sixty-tooth gear is actually moving one tooth to the next at a particularly designed rate. It is that escapement device that can be regulated to make sure that the gears are operating at the right speed. We could imagine some sort of braking device to slow the acceleration down to zero and have a continuous unwinding at a uniform rate, but then there would be nothing to count, no intervals. If that spinning could be broken, by stopping the wheel at equal intervals, we would have a device that counts the stops. Such a device is the ingenious gadget that was needed to build the first truly mechanical clocks.

From *The "Sphere" of Sacrobosco,* written in 1271, a popular astronomy and cosmographical textbook from the thirteenth to the seventeenth centuries, we learn that

> the method of making such a clock would be this, that a man make a disk of uniform weight in every part so far as could possibly be done. Then a lead weight should be hung from the axis of that wheel and this weight should move that wheel so that it would complete one revolution from sunrise to sunrise, minus as much time as about 1 degree rises according to an approximately correct estimate. For from sunrise to sunrise the whole equinoctial rises and about 1 degree more, though which degree the sun moves against the motion of the firmament in the course of a natural day.[11]

According to the twentieth-century British sinologist Joseph Needham, this passage gives some indication that the idea of a mechanical clock with weights and escapements was being thought about as early as the late thirteenth century. Needham claims that the engineering of such an escapement had not at that time been achieved, for he goes on to say that "the invention of the escapement seemed to have occurred at the beginning of the fourteenth century with no recognizable ante-period in a very advanced stage as regards design, though their workmanship was rough."[12]

The origins of almost every early clock invention are immersed in obscurity. Some historians put the earliest invention of mechanical clocks, with a distinction from clepsydrae, as early as the ninth century, while others put it as late as the fourteenth. Yet the evidence is sparse, and the records are scarce. Try to find the first pendulum clock, and you fall into the trap of what makes a clock a true pendulum clock. Even the invention of the pendulum itself is disputed; after all, it is just any hanging bob that swings back and forth. But a hanging bob that swings is also a clock of sorts that, like a metronome, could measure the second with extreme accuracy. The pendulum was known and used long before the earliest mechanical clocks were. We know that ancient astronomers in China used pendulums to measure something resembling time when observing and investigating eclipses. The number of swings could be counted that way with far greater accuracy than recording the shadow of a gnomon on a sundial, or the beat of a pulse, and far, far greater than the accuracy of a water clock. It didn't matter that the number of swings was not calibrated to normal sun time, for the fundamental property of the pendulum is that its frequency is stable under small arcs of movement. The time it takes for one swing at a small arc is extremely close to the time it takes for a swing at a slightly smaller or slightly larger arc. This fundamental property was well known long before Galileo rediscovered it.

---

Before the seventeenth century almost all clocks were large and expensive, usually funded by municipal governments to make sure that civilian goods and services ran smoothly. Clocks then were centered in town squares with hour and quarter-hour bells that could be heard for miles. Improvements—such as introducing a coiled spring, rather than a hanging weight, as the driving force—enabled clockmakers to miniaturize their works. But the basic idea remained: the transmission of energy from a source, such as a falling weight or an unwinding spring, to some oscillating motion that tracks the flow of time.

Clocks needed a driving force. Something had to keep them going, a hanging weight or a swinging pendulum feebly driven by an escapement. By the late eighteenth century, with the advantage of small, simple, and easily controlled coiled springs and miniature escapements, smaller clocks became possible. All clocks—aside from water, sand, and oil clocks—have, in principle, basic oscillatory generators. The best examples are clocks that work by stationary vertical springs and weights. By pulling down on the weight (thereby extending the spring) and letting go, the spring will oscillate up and down, losing energy to air resistance and heat through the molecular forces of expansion and contraction. By Hooke's law we know that when the spring is stretched or compressed, it exerts a restoring force proportional to the length of extension or compression. And if we were to consider quartz or atomic clocks, we could say that they too, in principle, work by oscillatory processes. For quartz it is the molecular and mechanical vibrations of the crystal, distorted by an electrical field, that vibrates at a frequency of 32,768 oscillations per second. For atomic clocks it is the natural oscillation of the cesium-133 atom, 9,192,631,770 oscillations per second.

All new inventions seem to start as big products before diminishing in size. My wife's first cell phone, a Motorola DynaTAC, weighed about two pounds and was as big as one of her winter boots. And so it was with clocks in the seventeenth century, when travelers had a particular concern for accurate time to coordinate arrivals, departures, and transfers of stagecoaches in a time when bandits were prevalent, and when weather could cause serious delays. Wealthy gentlemen often would carry heavy French carriage clocks in leather pouches when traveling. They couldn't see the time in the dark of the carriage, but they could press a button on the pouch that would connect to another button on the clock to ding away the latest time within fifteen minutes. In daylight they could flip open the top of the pouch to see the relatively more exact time. Pocket and pendant watches were available even in the sixteenth century, but they were impractically inaccurate, large, and mostly designed for curiosity, so the improved portable watches had to wait another century before waistcoat watches became available, yet still not mass produced.

By the mid-eighteenth century, clockmakers were able to miniaturize clock mechanisms so much that fairly accurate mantelpiece clocks were possible, and reasonably accurate pocket watches were made available to the few wealthy folks who could afford them—class symbols in the West. Clock design improved with a need for accuracy, and commercial demands of more and more accuracy continued with every civilization's advance of time, from the eras of global exploration and shipping to the Industrial Revolution to the computer age. It was a slow advance from one timepiece design to another. Pace kept up with need.

By the end of the nineteenth century almost every household in economically advanced western countries had at least one clock. Time was in control of everything one did; increasingly available precision was establishing a new order of behavioral regiments while dictating utilitarian routines and habits. Time was no longer casual, the fault of the more accurate clocks.

---

As late as the mid-nineteenth century, clocks were calibrated to noon at their local city. If you traveled from New York to Boston on the New Haven, Middletown, and Boston Railroad before 1884, you would have reset your pocket watch twelve times before reaching Boston. By the 1880s, railroad timetables were a chaotic mess of uncoordinated trains coming and going on crisscrossing tracks, sharing time, moving freight, and loading and unloading passengers at a limited number of station platforms. Precision of coordinated time was not a big concern before the mid-nineteenth century when—aside from the factory worker in the industrial cities who had to get to work on time—life was quite slow. Cows had to be milked, pigs had to be fed, the baker had to get up before the rooster, children had to get to school, and parishioners were expected to get to services reasonably on time after church bells called, but a quarter hour one way or another was as precise as one needed. All that was required was a central town clock that might just as well lose a few seconds each day as gain a few. If the town followed whatever time the clock said it was, the day would hum along as smoothly as the clock's hands. Precision was still of no concern.

In the 1820s the railroads came to Europe. For the next few decades they had short routes limited to travel between neighboring towns. By midcentury, trains were getting faster and the miles of tracks owned by

Boston Time Ball atop the Equitable Life Assurance Society building, Boston, 1881

one line were starting to connect with tracks owned by others. By 1860, European governments were seeing their ways into operating lines between neighboring countries. In the United States things were worse. Noon, defined in Washington, D.C., as 12:00 p.m., happened when New York was set at 12:12 p.m., when Sacramento's noon bell struck at 9:02 a.m., and when Boston time was 12:24 p.m. All across the country, noon was calibrated to the sun's position in the sky, so that Albany's noon happened two minutes after New York City's noon. There were no longitudinal time zones. Every city across the continent had its own noon.

On November 18, 1883, the *New York Times* published a column titled "Time's Backward Flight." It began, "Had there been stretched across the Continent yesterday a line of clocks extending from the extreme eastern point of Maine to the extreme western point on the Pacific coast, and had each clock sounded an alarm at the hour of noon, local time there would have been a continuous ringing from the east to the west lasting for 3¼ hours. At noon today all across the continent there will undoubtedly be considerable confusion."[13]

The day before, a Saturday, all across the country, people crowded jewelry shops to find out how to deal with the new, nonlocal time that was to be set on the next day. Many, hoodwinked by conspiracy prophecies, anticipated some unspecified disaster, a business catastrophe, or a more universal apocalyptic outcome. It was still a time when time balls fell from towers to mark noon as a calibration instant for local time.

Freight trains were using the same tracks as passenger trains, and the resulting crisscrossing of lines was rapidly expanding through government involvement. As sparring, expanding empires raced for shares of world trade, the need for a world time map became essential. At the 1884 International Meridian Conference, hosted in the United States, at Washington, D.C., twenty-six countries vied to permanently host a bench point location for mean solar time (12:00:00 noon).[14] They were to choose "a meridian to be employed as a common zero of longitude and standard of time reckoning throughout the world."[15] The Greenwich meridian was chosen as the home of the Prime Meridian. It is today considered the benchmark time for all human space travel.

---

Even after the new system was settled and habituated, there were those who believed that something was amiss—time stolen and time gained. It disturbed the comfort of solar time, when people knew that the sun was in control. Time didn't just jump. It got lost just a mile down a road. Suddenly, folks in Ober, Indiana, could walk across a road and lose an hour. Where did it go?

I felt the same way on a Friday flight to Japan from San Francisco. Crossing the International Date Line (IDL) at 10:27 a.m. Samoa Standard Time (SST), a rather arbitrary and imaginary zigzag that connects the north and south poles, my Friday suddenly became Saturday. At 10:28 a.m. Gilbert Island Time (GILT), I had lost all of Friday. Of course, I knew I had a credit of twenty-four hours in my time bank. It wasn't as if my lifespan would be one day shorter. Instead, as I pondered the notion that on return my credit would be cashed in, I also considered what would happen if I just continued around and around the world heading westward. Would my time credit increase? Would twice around give me two days' credit? Of course not! Flying westward is the counter direction to the earth's rotation, so the credit is being cashed in each time one of the twenty-three time zones is crossed until it reaches the twenty-fourth, when all the twenty-four hours of credit are used up. Pity.

Samuel P. Avery, engraver, “Universal Dial Plate or Times of All Nations,” in *Gleason’s Pictorial Drawing-Room Companion* (Boston), June 25, 1853, 416

Losing a day by crossing the IDL is an easy puzzle to solve, but it is one that easily leads to confusion. It appears as if governments tinkered with time when they agreed to divide time arbitrarily into zones. If one has little understanding of how clock time works, it is easy to be confused. By 1610 many countries in western Europe had adopted the Gregorian calendar. Britain, however, remained on the Julian calendar. Not until 1750, with the passage of the Calendar Act, did Britain—including its American colonies—bring dates in line with the rest of Western Europe. On September 2, 1752, the British government announced that there would be no September 3, 1752. The next date would be September 14: eleven days were eliminated. There was no confusion about the dates themselves, but some people might have thought that their lives would be shorter. Based on one of William Hogarth's satirical paintings, *An Election Entertainment,* which depicts a placard with a slogan, "Give us our eleven days," some history books reported that people rioted because they thought that they had lost eleven days of life and pay. Robert Poole, in his book *Time's Alteration,* seems to support some truth to this belief: "It was supposed by ignorant persons that the legislature had actually deprived them of eleven days of their existence. This ridiculous idea was finely exposed in Mr Hogarth's picture, where the mob were painted throwing up their hats, and crying out 'Give us back our eleven days.'"[16]

The frenzy was more myth than reality, since Hogarth was a satirical painter, not a historian. Moreover, the Calendar Act of 1750 was careful to anticipate misunderstandings by noting: "Nothing in this present Act contained shall extend or be construed to extend to accelerate or anticipate the time of payment of any rent or rents, or sum or sums of money whatsoever which shall become payable by virtue or in consequence of any custom, usage, lease, deed, writing, bond, note, contract, or other agreement whatsoever."[17]

Such tinkering with time, however sensible it may seem, plays with our intuitive sensitivity of time as the river that flows with continuous uniformity. It comes with a sensitivity coupled to the way we live and work, a feeling that time is absolute, that it behaves the same anywhere on earth, or even on the moon. Our twenty-four-hour day has a sacredness not to be tampered with. We've known it from a very young age and have lived with it for so long that any adjustments confound the way we live. Deep down, we know it is conventional. We have days, hours, minutes, and seconds. We go to sleep at some time, awake, work, eat, and play at other times, depending on either mood or circumstance. Our times are both arbitrary and conventional. But when we get used to them, they become elemental and so much a part of a system that they rouse our intuitions and stiffen the temporal components of our body rhythms.

Those arbitrary and conventional times with no universal significance give us our sense of time. It might simply be an illusion that something is out there, something like gravity or heat or atmospheric pressure, which moves things along in our lives to provide us with an absolute sense of time. If we were on the moon, measuring its slow turning of day to night, we might come up with a convention for time that is vastly different from our twenty-four-hour day. The moon rotates on its axis just about once in 648 hours. From earth, it seems that the moon does not rotate at all. That's because the moon is in almost synchronous rotation with the earth. In other words, the moon orbits the earth once every 27.322 days, and it also rotates on its axis once in approximately the same number of days. So, it appears to us that the moon keeps still. If, by convention, we divide a moon day into twenty-four moon hours, then each moon hour is twenty-seven earth hours, a very different time scorecard, and so a very different sense of time itself. And yet, although the scale of time is different—one earth hour being one twenty-seventh of a moon hour—our sense of time would not be much different, since we would remain the earth creatures that we are, tenaciously controlled by the commanding powers of our entrenched circadian rhythms.

Eugene Cernan and Harrison Schmitt, the Apollo 17 astronauts and last people to visit the moon (as of the printing of this book), spent twenty-two hours and four minutes maneuvering the lunar rover. To them it was an actual twenty-two hours and four minutes, but in moon time it was just about four-fifths of a moon hour. They were not sensing moon hours. Every consideration of time was geared to Greenwich Mean Time (GMT), even though they were on the moon. Any sense of time requires a sense of duration, not simply a mark of a particular time that is in no relation to another mark of time. The astronauts were busy with maneuvering the rover and collecting samples. They were aware of time's passing as a duration that began at the moment of starting the slow-moving rover and returning to the landing site. That was their time concern. So whatever time was marked from first setting foot in the moon dust and getting into the rover was simply a mark that would eventually be subtracted from the time marked when they returned with the rover—the duration. Duration is duration, no matter how it is measured. The sense of duration is, on the one hand, a human sense with all the experience and physiological rhythms tied to it and, on the other, a numerical measure that can be scaled to reflect that fitting human sense.

However, by that scaling, things get complicated. The scaled measure also depends on how deeply one concentrates on what one is doing during the duration. Imagine the intensity with which Cernan and Schmitt were absorbed in their work, looking for anything that might never have been seen before. Moon hours or earth hours—either way, their time must have been flying.

The hitch here is that we are talking only about a clock, not time itself. We have simply picked a periodic event and a counting mechanism for something we are calling time, as if it were simply attached to the cycle of earth's spin. Our twenty-four-hour clock, however conventional it is, is slowing down from hour to hour, being dragged by the moon and tides. Fortunately, we do now have an answer to the ancient question of whether the earth's rotation is constant.

For many years we've known that tidal friction is a drag on the earth's spin. But just how much the rate of rotation has varied was not known. Thanks to the work of Richard Stephenson and Leslie Morrison and their colleagues at Durham University in the United Kingdom in the 1980s, we have fairly precise measurements of how much the length of a day increases in a century. Stephenson and Morrison analyzed accounts of eclipse data dating back to 720 BC, including those found on Babylonian cuneiform clay tablets and in the Alexandrian astronomer Ptolemy's *Almagest*, along with reports from Chinese, European, and Arab astronomers. They then compared that data with computer models that mark the times and places where people would have eclipses *if* the earth's spin were constant from then until now. They found that in every century the length of a day increased by 1.8 milliseconds. So each day the planet spins a tiny bit slower than the last, making each day almost six hundred-millionths of a second longer.[18] We can blame it on the moon and the oceans dragged by the spinning earth.[19] So today is the longest day of your life so far, and tomorrow will be longer still.

## INTERLUDE: A CLOCKMAKER THINKS ABOUT TIME

Eighteenth-century Dutch clock with automata. Photo: Richard E. Bates

Ray Bates learned his clockworks craft some seventy years ago when he apprenticed to the renowned Edinburgh, Scotland, watchmaker R. L. Christie. After moving to Boston, and soon thereafter to Newfane, Vermont, he became known internationally as "The British Clockmaker," a restorer and conservator of fine seventeenth-, eighteenth-, and nineteenth-century clocks, aiming to preserve the integrity of the original maker. He tells us, "You have this relationship with the man who made the clock three hundred years ago, because you're touching the same metal that he worked on himself." I suspect that for Ray, that's the reward that comes from the hours of skilled work.

Ray is now retired, but his final apprentice is his son Richard, who has been working as an antiquarian horologist for twenty years. Richard now runs the family business. I ask Richard if there was ever a clock that inspired and thrilled him, perhaps a clock that deserved the splendors of its own inner workings. He tells me that he has just finished working on one, a Dutch musical clock with automata. "A very rare beast," he says. "It's a picture clock in a gilded frame. The picture is a pastoral scene, with cows in a meadow, a fisherman moving rods over a river, farmers scything on a hillside, windmills in the background, and a moon. But the moon is really a clock, and the entire structure is an automaton." He shows me a similar Dutch clock that he has been working on and then describes the automaton mechanism. "The cow bobs its head while eating the grass, the fisherman's line pulls a fish from the river every sixty seconds, the windmills turn their blades, horses bobbing their heads. There is an amazing ingenuity of pulleys and threads intermingling and indirectly connecting to the clock mechanism. Playing six tunes, all in Dutch, while the moon passes through its true phases and dials indicate the months and days of the week. Each month an exceptional engraving of a farmer appears in seasonal garments. So you can see why I'm taken with the Dutch clocks." Aside from the clock itself, the movements in the picture have no practical purpose other than to show off the skill. "I had the pleasure of taking that clock apart and repairing it. That was a thrill to work on."

I ask him how he thinks of time and what he thinks time might be. "For me," he says, "time is very concrete. As I'm working on a clock, I punch in the starting time on a time sheet, a punch clock that stamps the time. So, time for me is a measurement of activity. The clocks themselves can measure time in precise enough seconds for any reasonable human activity." He points to the clocks in his workshop and tells me that any one of them can count seconds accurately. Then, after pausing for a moment of quiet thought, he says, "Funny, I've been at this craft for over twenty years and never consciously, truly asked myself the question. I'm on the business letterhead and listed as a Horologist—loosely translated, 'one who studies time and the making of time pieces.' What I do here is restore these beautiful antique mechanisms in order that they might yet again provide the timekeeping qualities the original makers intended. We reanimate these handcrafted pieces, and for a moment we become one with the original maker of the clock."

It was not the answer I expected; rather, it was an openly spontaneous statement coming from someone who spends so much time working on mechanical clocks that he feels at one with their makers, a time traveling moment. The clockmakers would have given a similar answer. Whatever time may be in the abstract is not important. For the clockmaker, it's the measurement of time that is essential; it's the mechanics of time's movement forward that has ultimate value. All we need to know is the start to finish of a period of activity. One of the deepest insights of what time is comes from the art of knowing that time is passing, the interval of time spent, and the remaining interval of time before the activity ends. The clockmaker knows that, and knows that all attempts to nail down any deeper notions end in an illusion positioned in a phenomenological quagmire.

# 3

## EIGHTH DAY OF THE WEEK (CYCLES OF TIME)

On a flight from Minneapolis to Calgary, I overheard a child ask her parents, "Why are there seven days to a week?" The father answered, "It's because on the seventh day God rested from his work in creating the world." As I pondered the insufficient answer, the smart girl pressed on. "Yes, Daddy, but why seven? Couldn't God have taken seven days and rested on the eighth? Why did he choose six to finish?" The questioning came from a seat behind me. I was careful not to interfere, though I battled a strong impulse to turn and contribute my thoughts. Her question is the kind that children think about, the kind adults answer with indisputable, yet partial, understanding. The girl accepted the surefire biblical reference but latched onto the arbitrariness of seven. She kept asking, "Why not six or eight?" A flight attendant came by with drinks and snacks, suspending any acceptable answer the father might have had.

A potentially convincing answer—at least one that aims the father's answer in a less proverbial direction—comes from the time when Babylonian astronomers started examining the light of the moon and its four phases. I sat pondering the question: Could it be these four phases that give the seven days? Perhaps it all comes from the fourth of the Ten Commandments, "Remember the sabbath day, to keep it holy." The four phases are the waxing crescent, which starts with a new moon and lasts till the end of the first quarter; the waxing gibbous, from the end of the first quarter to the night of the full moon; the waning gibbous, from the night of the full moon to the end of the last quarter; and the waning crescent, which brings the moon back to new. Gibbous, in Latin *gibbous* (humped), refers to the light of the moon when it is larger than a semicircle and less than a circle. Each phase lasts about 7.4 days; perhaps that is where we get that magical number 7. The moon tells us that if we are to have any kind of a lunar calendar of repeating moon cycles every 29.53 days, then 7 is the closest whole number quotient when 29.53 is divided by 4. That number 7 is also the only prime number—evoking mathematically indestructible permanence—that sits in a practical spot along the string of possible whole numbers that could group days into a convenient cycle. That magical number 7 creeps in from other cultures as well, with 8 being a runner-up. It is coincidentally convenient that the earth happened to fall into an orbit taking about 365 rotations to go once around the sun. Equally convenient are the moon's frequency of phases. Divide 365 by the wonderful number 12, with its four divisors, to get fairly close to 30.4 days, conveniently close to 29.53 days, but closer to 8 when divided by 4. Imagine! Our weeks could just as well have 8 days. Perhaps it all comes down to the chance accident of how cosmic dust came to form our rotating planet four and a half billion orbits ago. Because of either God's design or wild accident, we have our days, nights, seasons, and, according to Ecclesiastes 3:1–4, a calendar of past, present, and future markings of our histories, appointments, predictions, and times for everything under heaven.

> To every thing there is a season, and a time to every purpose under the heaven: a time to be born, a time to die; a time to plant, and a time to pluck up that which is planted; a time to kill, and a time to heal; a time to break down, and a time to build up; a time to weep, and a time to laugh; a time to mourn, and a time to dance.

This Ecclesiastes list goes on, spanning a medley of diverse durations. A time for rest is surely one of purpose under heaven. Ignoring reason for how cosmic dust came to form our planet and put it into orbit, it might follow that a seventh day divides time into calendar weeks because in Genesis 2:2 we learn "that on it He rested from all His work." Dividing years into months, and months into weeks, puts milestones on the path of civilization and life. A calendar is a clock pointing to the you-are-here mark of the present day, yet it is more than just a clock with a twenty-four-hour cycle. A calendar pretends to cycle, but rather, as if in a third dimension, it helixes through eternity or, if not eternity, then till the end of humanity. It is a bookkeeper that provides a consistent ordering for the human body, the human mind, and the collective consciousness to continue to live and age together. Its function is to help us imagine the temporal scheme of past, present, and future as a schematic picture. Calendars precede the ordinary clock by thousands of years, and for good reason. It's hard to divide the day without markings from the sky, whereas it is easy to count days by marking nights. You don't need to be an astronomer to make a calendar, just a keen observer. If a clock is a schematic representation of time, a calendar is a schematic representation of ticking days, distinguishing one from another.

Early Egyptians had a disarrayed sense of the past because their calendar was ordered by the reigns of kings. Their culture had little concern for history, possibly because all glories would repeat themselves in other lives, although it did strongly recognize its history through art. The Jews, however, were more influenced by the Babylonians than by the Egyptians, and like the Babylonians, they based their calendar on the moon. In contrast to the Egyptians, the Jewish idea of time was based on both the past and the future, as an extension of God's purpose for the created world. Exodus tells the story of the Jewish people's freedom from Egyptian bondage in the second millennium BC, when they settled in Canaan, a region having both a fortunate and an unfortunate geography of being between Egypt and Babylonia. In 722 BC Assyrians invaded and destroyed the capital. Some Jews fled, some were deported to Assyria, and some settled in Jerusalem. Then, in 586 BC the Babylonians destroyed the Jerusalem Temple and deported many Judeans. I tell this story because it points to why Jews of that era thought of time differently than did their northern neighbors. They saw their misfortunes as punishments for having been unfaithful to God. Because they had so many misfortunes in their past, they looked to the future, when a messiah would come to rescue them, defeat their cruel enemies, and restore their civilization to its former glory. With the notion of a messiah, there is hope for the future, so time then had, and still has, a double focus for Jews. With Passover telling the story of Exodus, the past is kept alive. The Jews have their festivals that either recount the story of their people or sometimes, as with their Purim plays, reenact a story of the past. Time for Jews is not a one-way arrow but rather a symmetric feeling of double orientation. Not only is the past important for remembrance purposes, and hence significant for Jewish history, but also the future is believed to be critical in averting any repetitions of past torments.

Chinese dynasties based their thinking about civilization on the present, according to Alberto Castelli at China's Xiamen University. One would think that Buddhism, with its conception of afterlife and model of responsibility, would bring a forward sense of time to China rather than teaching us to be in the now. Though Buddhism came to China from India via the Silk Road in the third century BC, the idea of future was not as widely accepted there as it was in India. Modern Mandarin Chinese does not have a past or future tense, but it is a high-context language. In other words, to say something about the past or future, a Mandarin speaker either adds an external preposition or indicates through context that the verbs refer to something in the past or future.[1] For example, to say, "I went to New York," one would put in a time word, say, "yesterday." Literally, the sentence would come out as, "Yesterday I go to New York." The same applies for speaking about the future. To say, "I will go to New York," one might say, "Next year I go to New York."[2]

Unlike very early Indian, Persian, Babylonian, Hebrew, and Greek texts, Chinese culture has no creation story that starts with the creation of the world at a single moment in time. Chronologies start with the crowning of royalty or the transferal of dynasties, though not starting from any particular date. For centuries, biographies were presented as if the subjects were still alive, linked to "ancestral lineage, which includes both of the living and the dead."[3] Links between the living are strong in Chinese mythology, and so are mythical plans for the future. There is, for instance, the invisible red string of marriage, which binds two people who are destined for love. The origin stems from mythology involving the god of marriage and love, Yue Lao, who appears as the old man under the moon. Soon after the birth of a Chinese child, Yue Lao ties one end of an invisible red string to the baby's ankle. He ties the other end to the ankle of another baby of the opposite sex. Those two babies will meet someday and become betrothed. Yue Lao also predetermines the time of meeting.

The father of the child on my flight to Calgary had a safe answer to the child's question. While I was considering my own answer, guided by the mechanics of the universe, the father offered the time-honored deific one.

I suspect that, for him, Newton and Copernicus might have explained the mechanics of the universe, but orbits of moons and planets took their orders from God. God transcended all of nature, including the cosmos. As the first line of David's Psalm 19, praising the glory of God, says, "The heavens declare the glory of God; And the firmament showeth his handiwork." Unlike the Egyptians and Greeks, the Hebrews considered such creations as the sun, moon, and stars to be God's handiwork.[4] Calendars of seven-day weeks, however, appeared almost a thousand years before King David's time, when Babylonians worshiped as their patron deity Marduk, the god of water, vegetation, and judgment, as well as the god of supremacy in the present, when calendars were used mostly to mark festivals of either coronations or religion. As the British mathematician and science historian Gerald James Whitrow pointed out, "The king-priest was the incarnation of the invisible god in the sky and the rituals he performed were the repetition of divine actions and had to correspond exactly in time as well as in character with the rituals on high."[5]

Hellenistic time had more of a cyclic pattern. Stoic cosmology saw the world as a collection of worlds "that had been" and will repeat. There was always something mystical when the sun, moon, and planets would line up in the sky. Even back in Babylonian times, such a year was called "The Great Year" to enhance its sacred prominence. We can easily understand that such a rare phenomenon as the lining up of sun, moon, and planets was so special that for those inclined to believe in signs, a Great Year would signify an impending colossal event. Just think about it: the sun, moon, and planets appear as a kind of cosmic clock whose hands take a long time to repeat, yet every once in a long while they do. If the heavens return to where they once were some time ago, then everything else that depended on the heavens should, by reasonable association, restore itself to what it was before. The Great Year was described by a fourth-century bishop of the town of Emesa (present-day Homs) in western Syria as follows:

> The Stoics say that when the planets return, at certain fixed periods of time, to the same relative positions, in length and breadth, which they had at the beginning, when the kosmos was first constituted, this produces the conflagration and destruction of every-

> thing which exists. Then again the kosmos is restored anew in a precisely similar arrangement as before. The stars again move in their orbits, each performing its revolution in the former period, without variation. Socrates and Plato and each individual man will live again, with the same friends and fellow-citizens. They will go through the same experience and the same activities. Every city and village and field will be restored, just as it was.[6]

This lineup is also clearly told in Plato's dialogue *Timaeus,* where we are told that "there is no difficulty in seeing that the perfect number of time fulfills the perfect year [The Great Year] when all the eight revolutions, having their relative degrees of swiftness, are accomplished together and attain their completion at the same time, measured by the rotation of the same and equally moving."[7]

Alfred Russel Wallace, a British explorer and an evolutionary biologist who published papers with Charles Darwin, wrote that since there was (at that time) evidence to believe the earth to be the only inhabited planet in our solar system and likely the only place in the whole universe that ended up adapting the conditions to support life, there is indication that the whole universe was "precisely adapted in every detail for the orderly development of organic life culminating in man."[8] From his writing, it is not at all clear if he is suggesting that the universe was created with some divine goal to produce life "culminating in man," yet Mark Twain took it that way when he wrote his rebuttal quip to Wallace:

> Man has been here 32,000 years. That it took a hundred million years to prepare the world for him is proof that that is what it was done for. I suppose it is. I dunno. If the Eiffel Tower were now representing the world's age, the skin of paint on the pinnacle-knob at its summit would represent man's share of that age; and anybody would perceive that that skin was what the tower was built for. I reckon they would, I dunno.[9]

If the spin and orbit of the earth were divinely inspired, they could not have been better intended. I feel fortunate to have been born on this planet with a seven-day week. Could we live with an eight-day week? I dunno.

# PART

# THEORISTS, THINKERS, AND OPINIONS

If the greatest philosophers became embarrassed when they investigated the nature of time, if some of them were altogether unable to comprehend what time really was, and if even Galenus declared time to be something divine and incomprehensible, what can be expected of those who do not regard the nature of things?
—*Maimonides,* The Guide for the Perplexed

# 4

## ZENO'S QUIVER (THE STREAM OF TIME)

Almost every child is born a philosopher. Children ask questions that professional philosophers spend lifetimes brooding over. When not playing stickball, handball, or flipping cards summer afternoons in the Bronx, I'd be with my little band of friends on someone's front stoop, debating such unanswerable questions as *What place would the Yankees be in if Mickey Mantle had never been born?* or *What's the biggest number?* or *What is the smallest thing in the world?* We didn't question time, but that smallest thing was a regular on our stoops. We'd take turns tearing matter down to atoms, and atoms to electrons, before getting to the annoying, "Yeah, yeah, an electron, but take an electron and break it up. Don't ya get somethin' smaller? I mean it's gotta be made of somethin'!" For a ten-year-old, it was very disturbing to think that a piece of something could not be broken down into something smaller. Kids all over the world think about that kind of question. That's adult philosophy, asking questions that have answers we tend to take for granted. Of course, we had no thoughts of quantum theories of space, where Planck length is believed to be the shortest meaningful length in the universe (less than $5.3 \times 10^{-35}$ feet), a length below which any attempt to measure physical size would run into total uncertainty and nonquantum usefulness.

Questioning time has almost always been central to philosophy, along with questions of whether or not time exists outside our own thoughts about it and whether or not time exists beyond the present moment, that moment that is gone the instant it arrives, that strict, inexpressible point that theoretically divides the past and future. But such a point exists only in the mind, so the only real way to think of it is by loosening the strictness and envisioning an interval, a slightly longer moment that is deceptively conceptual.[1]

The pre-Socratic philosopher Parmenides of Elea questioned time in the early fifth century BC. His student Zeno, also from Elea, spoke of time as a string of minuscule moments, concluding the nonsensical impression that motion is impossible. Though the rules of logic had not yet been established, his flying arrow paradox, for instance, seemed logically to conclude that it is impossible for a thing to be moving during a period of time, because it is impossible for it to be moving at any particular indivisible instant. An arrow shot from a bow must be in some place along its trajectory at any instant you pick. Being in that place and time, it must be stationary. In other words, whenever you look at the arrow, it is stationary. How, then, does it get from one place to another? Since it never seems to be moving, it can never reach its target. Indeed, it cannot even leave the bow! In Zeno's mind, looking at the arrow while freezing time at any instant was like taking a picture, a picture of an arrow not moving. We know, though, that that picture—if by picture we mean the modern word *photograph*—was taken, not at an infinitesimal instant, but rather taken *along* an imperceptibly short interval of time that had a beginning and an end. We know that that picture is microscopically a blur. But Zeno didn't see it that way. He would insist that at any particular time the arrow must be someplace, thereby leading to a paradox of time and motion. That arrow paradox, along with others of Zeno's, raised a fundamental question of whether time and space are continuous or perhaps come in discrete units, like a string of beads.

Zeno's argument asked us to stop the arrow in flight, to stop time so that we might examine a stationary arrow without destroying its flight. The mathematician can do that easily—stop time to visualize the arrow abstractly—and to believe that the frozen arrow is indeed one and the same as the one shot. But he or she is simply replacing a mathematical abstraction with a mental impression of a fixed arrow—one that may even be clearly visualized as if on a screen in the mind; it is not the real arrow that smoothly advances on trajectory to its target.

We know that the arrow flies through the air; we can see it happen. Yet there is difficulty in explaining why or how we know. In mathematics and physics, time is a variable that can be fixed by simply declaring it to be some number. We have formulas that tell us where the arrow is at any time *t*, so if we let *t* equal some specific time, then we should know the exact spot where the arrow is at that time. Yet this means that our mathematical models of motion, space, and time are intellectual constructions built for the convenience of easy calculations—and also, we hope, for the greater purpose of representing the structure of reality. And that structure should assume that we know what *t* actually is, besides being just a number, or at least that we know how it relates to what we think we know as time.

Almost all of what we know about Zeno's life is speculation, composed from fragments and historical sources written almost a thousand years after his death. We know that he wrote a magnificent book on philosophy that was used as a textbook at Plato's Academy, but not even the smallest fragment of it has survived. The fifth-century philosopher and mathematician Proclus, our principal source of information about the early history of Greek geometry, tells us that Zeno wrote a book containing forty paradoxes but that it was stolen before it could be published. Four acknowledged paradoxes come to us by way of Aristotle alone. Dozens of major works written by renowned scholars from Plato to Bertrand Russell have pondered them to fill tomes of literature containing grand connections arching across history.

The absence of Zeno's writings warrants suspicion over whether the man actually existed beyond merely being a character in Plato's *Parmenides*. Despite that absence, a great deal of extant material tells of his profound philosophical ideas; one can gather enough from them to assemble a coherent story, whether the man lived or not. Plato and Diogenes Laertius provide the corners to the jigsaw puzzle of Zeno's life while Aristotle and Proclus give the edges of his philosophy.[2] We fill in the rest with supposition.

Each paradox in Zeno's quiver showed continuity as merely a conscious impression, a fabrication of the mind elevating illusion to reality. Recall that fabled race between the swift-footed Achilles and the slow-moving tortoise. Before Achilles gave the tortoise a head start, he should have known that he would at every instant be catching up to where the tortoise once was and therefore that he was doomed to lose the race. Of course Achilles would lose if we model as an illusion that the race is simply a matter of moments of catching up.

Though mathematicians may try to explain the paradoxes by logical models of motion phenomena, using algebra or infinite series, they miss the target—to give a phenomenological explanation of the unavoidable sense of harmony between an illusion of time and the continuously flowing universe. Yes, they can tell us precisely where the arrow is, when Achilles will overtake the tortoise, or when we will come to the other side of a room, but they cannot tell us why without bending our perception of space to fit our inflexible intuition of time's continuous nature. Achilles overtakes the tortoise because we can work it out through algebra based on the continuous nature of the real number line. However, we cannot imitate precisely any phenomenological nature of real matter that is composed of atoms with their excited electrons changing their orbits only by discrete jumps and their energies changing by discontinuous quanta packets.

Zeno conjured us into thinking that his arrow moved in discrete jumps, that Achilles was always catching up to where the tortoise once was, that we move across a room in an infinite repetition of moves of half the distance we intend to cover. It took us a while to realize that another way to approach the paradox of motion is to ponder the paradox of measuring time, hence to conceive the framework of numbers.

---

"What is the wisest thing?" Pythagoras rhetorically asked. "Number" was his expected answer.[3] That wisdom has been repeated for thousands of years. We know he was right; perhaps not everything, but almost everything, comes down to numbers. In this digital age we find that even the pictures on our smartphones are entirely made from numbers. And the blue sky on a sunny day? That too is made from numbers.

A blue sky is generally seen as a shading of blue, a gradation of light by wavelengths that change in the range of 450–95 nanometers (nm) as a function of position angle in the sky. A nanometer is one billionth (0.000000001) of a meter. Blue, at wavelength 451 nm, is indistinguishable to the human eye from blue at wavelength 450 nm. The blue sky is painted with blues by a wide range of wavelengths. There is a next-ness, because wavelengths are always discrete whole numbers, but that next-ness is imperceptible. Imperceptible, yet measurable by its wavelength. It suggests that though the position angle has no next-ness, the color can change only by discrete wavelengths. It confuses the difference between mathematical continuity and real world continuity. It comes down to the difference between mathematical modeling of the tiniest elements, gaps, and moments and the real world of atoms and the excited behavior of their subatomic parts. The clock can be an astoundingly precise measuring instrument of time; however, time itself has that fascinatingly subtle glory of being able stealthily to elude the most precise clock in the world. By convention, a second is now 9,192,631,770 radiation periods of cesium-133. In cesium-133 ($^{133}Cs$), electrons move from one state to a higher one, radiating photons at a stable frequency of exactly 9,192,631,770 Hertz. Atomic clocks can now measure time with astonishing accuracy, counting those radiating photons of cesium-133. Even they are off by just short of a second every twenty million years. And even those clocks can never quite catch the precision of time, for they measure time by counting through a line of whole numbers, each followed by the next while time itself moves rapidly through a blur of real numbers, numbers with no neighbors that are "next" in their stream of order.

Before 1956 the second was ephemeris, defined by the earth's rotation on its axis or its orbit around the sun. It was an unstable definition, since the moon's gravity slows the earth's spin. When astronomers from the U.S. Navy and the British National Physics Laboratory researched the frequencies of the metallic chemical element cesium-133 oscillations, they found that those frequencies are far more constant than the frequencies of earth's orbit around the sun. We generally relate speed to clocks that give us a measure of change, but change does not necessarily need a clock to give us those measurements. We might say that a train is moving at a hundred miles per hour. But if we take a baseline of, say, a single beat of the human heart, we could simply correspond the speed of the train to our base. The result is that the train's speed is approximately 450 human heartbeats. Heartbeats, not a ratio of two mismatched units. Time in that scheme is not at all in the relationship. But let's not be fooled. Time is there. We just substituted the heart for the clock. The units are different, yet both heart and clock tick away at their own paces.

As a clock moves through a sequence of escapement swings, or even cesium-133 oscillations of energy levels, it is skipping over relatively large segments of time that are measured without being marked. This is a meeting of the real world with the world of nontactical ideas that are far beyond what we can mark and measure precisely with nonmathematical a real point. Like the now, it too is a moving target that is so thin in the number line that it can never, ever be hit with an infinitesimally small pointed dart. The now moves so fast, relatively, that you only know it when it's already in the past.

To give a clearer picture of the distinction between mathematical continuity and real world continuity, it is essential to go back to the time of the fifth-century BC atomists Leucippus and Democritus, whose school of natural philosophy believed that all material things are composed of *ἄτομον,* indivisible bits and pieces. According to this ancient theory, atoms are indestructible, indivisible bits of things of various forms and sizes that could fit together in bonds and collisions in the void to form the bigger pieces of stuff in the macroscopic world. It was a theory that came naturally into metaphysical thinking; after all, whenever a curious person thinks about what a thing is composed of, no matter how informed the person is, the atomism thought quickly surfaces. Surely my little band of Bronx friends thought that they had discovered the atomist idea.

To imagine the tiniest elements of time, we must appreciate somehow what is known about infinitesimals. The story of infinitesimals goes all the way back to the Delphic problem, a story involving the oracle of Delphi. When Apollo had set a plague upon the island of Delos, the oracle was consulted on what to do. She responded that the black marble cubical altar to Apollo must be doubled in volume and yet remain a regular cube. It was a peculiar prophecy with a perplexing meaning. The Delians carved a new altar, misinterpreting the mathematical meaning, by doubling all four sides. To their surprise, the new altar ended up being a solid cube eight times as large in volume.

Plato was consulted. He explained that the oracle's advice was a warning from Apollo that Greeks were neglecting their study of geometry.

Doubling the volume of a cube was at that time a rather profound proportion problem involving some tricky geometry. But Apollo might have had a weightier motive: to stir "the entire Greek nation to give up war and its miseries and cultivate the Muses, and by calming their passions through the practice of discussion and study of mathematics."[4] The original story was relayed to us by the Greek philosopher Theon of Smyrna from the third-century BC Platonic philosopher Eratosthenes, through his book *The Platonist*.[5] It involves the ancients' futile attempts at solving this problem, one that seemed simple: given the edge of a cube, construct the edge of a second cube whose volume is double that of the first. The doubling problem, however, must be solved with just the tools of straightedge and compass, because to prove its success the only accessible logical tool would have to have come from Euclid's first principles.

The algebraic equivalent of duplicating the cube is finding the cube root of 2. Note that doubling a cube whose side has a length in units means constructing a cube whose side is of length $\sqrt[3]{2}$. However, in fourth-century BC mathematics, numbers had to be rational, ratios of whole numbers. Because the cube root of 2 is not rational, the tape measure that measured an edge of the cube (which, after all, would measure a side as one unit) could not also measure the diagonal. Irrationality of the cube root of 2 was well known in the fourth century BC. And because irrational numbers were not known as numbers then, it was believed that there must be gaps in the tape measure, missing numbers. This in turn suggested that there are things in the world that cannot be measured, an uncertainty principle that must have been a startling revelation. It must have been quite a blow for natural philosophy. In effect, the oracle was concerned with a mathematical understanding of those gaps, those missing numbers.

How does Zeno's moving arrow, which after all is moving through points represented spatially by real numbers of the number line, get to the next point along its trajectory when there is no next point? A notion of next point is meaningless in the geometry of the real (or even the rational) number line. For instance, take the rational number ½, which, in decimal notation is 0.5. What is the next rational number? It cannot be 0.51, or 0.501, or 0.5001, or any number starting with a decimal representation of ½ ending with some long string of 0s with a 1 at the end, for such a number would be farther from 0.5 than one gotten by slipping in another 0 before the final 1. So, if the tip of the arrow has traveled, say, ½ its anticipated distance, where does it go next? It seems as if the arrow flies along its path discontinuously, in and out of its own existence.

---

Aristotle disputed Zeno's arguments in his *Physics*.[6] One argument for the connection between mathematical continuity and real world continuity comes from observing that a traveling object cannot skip positions—it must move from one position to the next. Aristotle was a rationalist, not an atomist. For him, the continuity of space did not imply the infinite division of the object traveling through space. This does seem contradictory, for how can an object move from one position to the next without space coming in discrete units—that is, without a position that we might consider as a next place? According to Aristotle, "Time is not made up of atomic 'nows,' any more than any other magnitude is made up of atomic elements."[7]

It suggests that movement can happen only by a moving agent. A stone carver's chisel cuts the stone, the potter's hands shape the clay, and the weaver rapidly pushes the weft and shuttle back and forth across a warp through a perfectly synchronized opening and closing heddle shaft. For Aristotle, a mover must be in direct touch with the thing it immediately moves. He could not have known about rods and cones on the retina, yet his direct touch argument for many cases of nature seemed to work with hearing and seeing. Hearing happens when air particles hit eardrums. Seeing happens when light waves stimulate retinas. Emotions, such as fear, anger, and love, he attributed to blood flow, and claimed anger to be a kind of furious frothing of the body's blood or an overheating of the heart. For him, mind was in the heart and the eyes were windows to the soul. And all things could be explained by one thing touching and moving another in some span of time, as if the whole world worked by a system of gears and pulleys. It is the sort of belief that my preteen coterie of philosophers on a Bronx stoop would have accepted, had it not been for discoveries of the magic superpowers evoked by Superman comic books of the 1950s.

But just as motion needs time, time needs motion. It seems that time is the measure of motion and vice versa. The noun, *time,* is joined to the verb, *to time.* In order to have a thing to call time, there must be something to time. In his *Physics,* Aristotle wrote, "So, just as there would be no time if there were no distinction between this 'now' and that 'now,' but it was always the same 'now'; in the same way there appears to be no time between two 'nows' when we fail to distinguish between them." So time and motion are inseparable. He asks us to try to imagine time without movement or movement without time. It's impossible. "Even if it were dark and we were conscious of no bodily sensations, but something were 'going on' in our minds, we should, from that very experience, recognize the passage of time."[8]

Therefore, if time is continuous, so is space. Yet time is divided by this curious thing we know as "now"; and, by the same reasoning, so is space. The position of any object in motion is marked and divided by its "now" place in space. But that does not exclude the concept of a smallest unit of time or space. Like my young Bronx friends, Aristotle surely understood that an interval could be infinitely divided, but his conception of infinity grants that we can always imagine a "beyond"—that is, a potential for continuing indefinitely and also that our minds have the power to continue to divide a line or an interval of time as often as we like. But those divisions refer to rational numbers. So he used this potential infinity to argue that Zeno's dichotomy paradox was based on the false belief that it is impossible for a thing to take up an infinite number of positions in a finite amount of time. That paradox maintains that a moving object will never reach a given point because however near it may be, it must always first accomplish a halfway stage, and then the halfway stage of what is left, and so on. Because such a series has no end, the object can never reach the end of any given distance. In effect, a moving object would have to "count" infinitely many numbers before the end of its journey.

That line of reasoning has a convenient loophole that makes it possible to perform an infinite number of tasks in a finite amount of time. The German mathematician David Hilbert's famous infinite hotel trick is a model example: Somewhere in math-wonderland there is a hotel with an infinity of rooms numbered 1, 2, 3, . . . The hotel is always full, but there is always room for one more guest. The manager tells the occupants of room 1 to move to room 2, the occupants of room 2 to room 3, and so on. This frees up room 1 for the new arrival. This annoying inconvenience for the guests may seem impossible to accomplish in a finite amount of time, given that each guest must move in real space and real time. But if the occupant in the first room takes ½ hour to move, the occupant in room 2 takes ¼ hour, and the occupant in the *n*-th room takes $\frac{1}{2n}$ hour, then the infinity of moves will be finished in just one hour! Of course, this assumes that the guests will be increasingly speedy.

By that, it seems Aristotle had made false assumptions in asserting that it is impossible for a thing to take up an infinite number of positions in a finite amount of time. Aristotle points out that time and space are equally divisible without limit, and therefore there should be not be any surprise that a person can pass through an infinite number of positions in a finite amount of time. But there is more to his refutation of Zeno's dichotomy paradox. Zeno must have assumed that when the path of motion is bisected, the motion is interrupted; the bisected point would then be considered twice—once at the end of the first segment and again at the beginning of the next.

With any assertion that movement happens continuously, it follows that as a body moves, there is no jumping into and out of its own existence. And from that it also follows that something undergoing change cannot change from *here* to *there* or from *this* to *that* all at once; otherwise, there would have to be some instant when the whole thing was neither *here* nor *there*, neither *this* nor *that*. This is reminiscent of another part of Zeno's philosophy, which argued that anything that changes must change in time and space, for anything that cannot be divided in time cannot be made to move in space. These arguments forcefully lock up the connection between time and change, and time and space, suggesting that time is the measure of motion—and that motion is the measure of time. So, in a sense, time *is* motion. By motion we don't mean just *locomotion*—the movement of an object from one place to another. Motion can mean a change in quality (as in change in color) or quantity (growth or shrinkage in size). It can mean the ripening of a pear or the bonding of atoms.

---

If we were to ask Zeno why we see the arrow leave the bow and hit its target, he would still respond, "Mere appearance of change. Motion is an illusion," and might add, "Now that you've had more than twenty-five centuries to ponder the problem, you know that even matter is nothing more than energy and vice-versa. Nothing has changed. The external world may be made from material known only by our senses giving the illusion of color, smell, feeling, and motion." My guess is that he would maintain that time is also an illusion.

The twentieth century brought us relativity and quantum mechanics. Space and time were no longer thought of as separate aspects of reality; they were united into a single four-dimensional continuum. Time dilation, inconstancy of mass, and special relativity suggest that motion is indeed illusory. Quantum theory suggests that some motion is not continuous and therefore that time is not continuous. Electrons are strictly confined to moving between discrete energy levels around an atom's nucleus. It's hard to imagine them discretely jumping around, for it disrespects our sense of continuous time. Surely, Zeno would rejoice in learning that his quiver of paradoxes cannot be cast off by simple calculus arguments.

Zeno's arguments remind us that time might not be as continuous as it seems. Is there an elemental unit of time that cannot be split? Could time, like light, be composed of minuscule particles, quanta? The German mastermind of quantum mechanics Werner Heisenberg once suggested that the smallest unit of time is something in the neighborhood of $10^{-26}$ seconds. And yet there is also Planck time ($5.39 \times 10^{-44}$ part of a second) named after Max Planck the German quantum theorist, a unit of time by which light travels in a vacuum one Planck length (roughly $6.3 \times 10^{-34}$ inch), and by that any time unit smaller would have a negligible effect on any nonquantum special existence.

Those mysteries of time, with all their technological and scientific advances, lead to some of the greatest questions of our civilization. Zeno conjured us into thinking that his arrow moved in discrete jumps, that Achilles was always catching up to where the tortoise once was, that we move across a room by considering an infinite sequence of distances, each one half the one before. The paradox of motion is the paradox of measuring time.

Physicists have introduced a theory called "loop quantum gravitation," a unification of quantum physics with gravity based on the possibility that space and time are composed of discrete pieces. If there are subatomic particles of matter, then why not have time composed of subtemporal specks? On balance, space-time itself might be built from specks of space-time dust, whatever they might be. Loop quantum gravitation theory is based on an interesting vision of the almost infinitesimal local structure of space, a structure that envisions space as granular. It comes from combining what is known about quantum mechanics and what is known about general relativity. It predicts that space itself has its building blocks at the size of about $10^{-99}$ cubic centimeters and that time moves in discrete jumps of $10^{-43}$ second.[9]

Time quantized? Why not? Like the quanta of light that we call photons, there could also be quanta of time that we might wish to call *chronons.* After all, photons are granular, and atoms have discrete energy levels. Our natural world experience gives an impression that space and time are continuous, so it seems hard to conceive that they aren't so. Unfortunately (or perhaps fortunately, depending on one's confirmation bias), loop quantum gravitation theory has not yet been able to make predictions already confirmed by general relativity, so it remains a theory, like string theory. String theory also gives us a picture of an extremely small and discrete world of space and time far beyond the reach of experimental observation, even by the most sophisticated experiments of particle accelerators that deal with magnitudes of specks of space smaller than $10^{-20}$ the size of an atom's nucleus.[10]

It is reasonable to expect discontinuities in time. First, there is the argument that if time is a part of a discontinuous space-time fabric, then time itself must also be discontinuous. But that argument suggests that if the fabric has gaps in one dimension, it must also have gaps as a whole, and that is not necessarily so. A second argument is more reasonable. Any measure of time must be connected to some kind of counting principle, a counting of ticks and tocks, whether it is the counting of waves that signal electrons changing energy levels in some isotope of cesium, the counting of ratchet releases in an escapement, or the counting of swings of a pendulum. Even the volume of water at the base of a water clock is a counting of special mass and therefore at some microlevel is discontinuous. So, we might have to face the bizarre thought that space is a flickering movie through time. And even though we still use metaphorical wordings such as *stream of time* for qualitative purposes, time passes more like an hourglass of falling sand than a continuous stream, whatever that is. That being the case, we might wonder what happens to the existence of the world between those falling bits of time sand. Like the workings of a movie: images pass through a gate, one at a time, disappearing before another appears to give the impression of continuity. Existence could be just that kind of illusion.

> It passes like the frames of film
> Through an aperture of still nows
> Appearing as lapsed realness
> Of time's onward illusions
> Driven by a ticking tailwind
> All the way from Big Bang to date
> Dropping droplets of memories
> Shedding shadows of history.
> —*JM*

# INTERLUDE: PRISON FOR LIFE WITHOUT PAROLE

What is time like for somebody crossing off the days of life in prison? Jason Hernandez was just twenty-one years old in 1999 when he was sentenced to life without parole. One cannot imagine how time ceased to exist for Hernandez. In 1981 the Dutch composer Louis Andriessen composed an orchestral work called *De Tijd* (The time) in which the listener is meant to feel as if time ceases to exist.[1] It was inspired by Saint Augustine's speculation, "If only men's minds could be seized and held still! They would now see how eternity, in which there is neither past nor future, determines both past and future time."[2] Hernandez was living in the eternity of ceased time.

"In my neighborhood," he told me, "you either sell drugs or you use drugs. You drop out of school, you go to prison, you come out—I just didn't see anything wrong with it at the time."

Caught, and charged with eleven counts of conspiracy to possess with intent to distribute controlled substances, in particular cocaine, Hernandez was sent to a high-security concrete and iron prison surrounded by twenty-five-foot walls and barbed wire in Beaumont, Texas.

"It sends you a message. You don't think about trees or rabbits. There's nothing past the barbed wire, 'cause it's not there for you anymore. You're caged like an animal, so you start acting like one.

"I never knew or imagined how the time would pass. I just knew it would. So I was basically preparing for a day, a moment in time that I did not know what it would hold for me or in what way it would happen, but

I knew it would happen. . . . I did wrong. I deserved to be punished, but I don't think I deserved a life without parole."

The natural human survival psyche gives such hope, so it was healthy for Hernandez to think of time moving forward with hope from the day he was convicted. For him, it wasn't hope that encouraged him but rather faith that he would get out. That faith was his survival in a sentence with no end date.

"To speed up time I tried to create an illusion that I still had a life and still had some importance. I had a routine of waking at 4:30 every morning and going to sleep at 11:00, but sleep in a prison is never continuous. Routine and trying to feel like you have control in a place where you have absolutely none is hard. I would wake up three or four times a night. Some guards like to clang their keys as they pass cells, just to let you know that they have the keys."

Then came the shock of learning that his brother was murdered in prison. That moment changed the way Hernandez imagined his future. Fighting for his freedom, he saw himself as a supporter of inmate rights, as an advocate of fairer sentences for nonviolent crimes. He saw himself achieving a doubtful role of speaking with lawmakers in legislatures and courthouses about unjustifiable, nonviolent crime sentences. And so, while in prison, Jason Hernandez studied federal law, wrote appeals without the help of an attorney, and advised other inmates about law and appeals.

"I became the prison attorney, which gave me value and a sense of importance. I became a casino operator with a poker table, a blackjack table, and a betting line for sports. I ran a grocery store selling commissary for a price and a half—Cokes, candies, chips, and so on. I ran a restaurant that sold nacho bowls, pizzas, burritos, and tacos. I felt that time would go faster if I kept myself occupied, but it also gave me social status and a sense of importance."

He knew that President Barack Obama had already received 33,149 petitions for commutations, and yet, with his faith in the system, he submitted an eight-page presidential clemency application that he got from the prison library. He imagined that Obama didn't get many clemency letters; so, just to add a dash of soul compassion to his chances, he wrote a personal letter directly to President Obama, asking for clemency. Yet time moved on in strange ways, as when he was blamed for the stabbing of another inmate and sent to the SHU (solitary confinement).

"The SHU is an unimaginably dehumanizing place with small strips of light through a window with blinds. You can't see out. The walls are concrete and the only things in it are a shower, a bed, and rats. At first, I plugged up the rat holes with socks and towels. I had no human contact for almost twenty-four hours a day." But then, within a few days, after being overtaken by the first of many body-leaving hallucinations, he longed for the pleasant company of a rat or two. "When you are sent to solitary, you are not told the length of your sentence. That's part of the torment behind the sentence—it could be a week, a month, or two years. I had mates who had been in the SHU for more than a year." So every day in the SHU was severe agony of timelessness, a torture that fired wildly in his brain.[3]

At the end of our short chat, Hernandez told me that his entire experience, that seventeen-year span, was a small price to pay for the freedom and opportunities he now has.

In 2015 President Obama commuted Hernandez's sentence to twenty years.

# 5

## THE MATERIAL UNIVERSE (PHILOSOPHERS' TIME)

For what is time? Who can easily and briefly explain it? Who even in thought can comprehend it, even to the pronouncing of a word concerning it? But what in speaking do we refer to more familiarly and knowingly than time? And certainly we understand when we speak of it; we understand also when we hear it spoken of by another. What, then, is time? If no one asks me, I know; if I wish to explain to him who asks, I know not.
—*Saint Augustine,* Confessions

I suspect that almost every adult philosopher, at one time or another, asked questions comparable to Augustine's "What, then, is time?" The question might be too perplexing to yield a fathomable answer. We don't even have a simple answer to how time began. Plato gave us an account of how time began with the creation of the universe. His fictional character Timaeus suggested that time never has been and never will be anything other than what the sun, moon, and stars tell us it is. Timaeus is portrayed as a philosopher-physicist, hence a person of trust in reason, and therefore a person who knows what he is saying when he tells us that we can talk about the past and future only as a moving image of our shared passing through eternity: "Wherefore he resolved to have a moving image of eternity, and when he set in order the heaven, he made this image eternal but moving according to number, while eternity itself rests in unity, and this image we call time. For there were no days and nights and months and years before the heaven was created, but when he constructed the heaven he created them also."[1]

So, in Plato's view, the cosmological clock's dials move with all the motions of everything that moves, as if all movements mesh in some kind of connected gear exchange to give us time. In other words, time *is* the movement of the heavens. The birth of time was the birth of the heavens and the "moving image of eternity": "Time, then, and the heaven came into being at the same instant in order that, having been created together, if ever there was to be a dissolution of them, they might be dissolved together."[2]

It makes sense. Our normal naive perceptions of time come from our experiences with daylight. We tend to think that there is something out there in the universe of invisible agents that dictates some universal notion of time, some intangible clock that beats out the pulse of the universe. The earth makes a full rotation in twenty-four hours, and it takes a year for it to complete its cycle around the sun, by the definition of *hour* and *year*. If it were not for contemporary physicists telling us otherwise, we would imagine time as a master gauge of the universe, moving at some precise, uniform rate. Any clocks that we have, our pulses, the orbits of our sun, planets, and moon, would move along in calibration to the grand master gauge of a universe tolling to the rhythm of absolute time. Of course, we can use the sun's orbit as a clock; one complete cycle marks a unit of time, which we might call a year, but that is just a timepiece calibrated with what might be absolute time, if there were such a thing. All durations would be measured against that imagined grand clock in the sky. So, the moon orbits around earth in one interval, the earth around the sun at another, and the dial of a watch at a third. All these intervals can be compared, but all would be secondary to the absolute that presumably had always been, had there ever been an absolute.

An absolute suggests that time started from zero. But a time when time did not exist seems unfathomable. After all, the Big Bang had to have happened at some time! If cosmologists tell us that it happened 13.7 billion years ago, time must have been moving forward ever since; otherwise, what could that "13.7 billion years ago" mean? After the Big Bang, for the next almost 400,000 years, there was not much more in the universe than a googolplex of atoms wearily trying to form something as simple as hydrogen, a gas needed to form the first star. There were no stars. Where and what was time before the first galaxy appeared 400 million years later?

---

Now, a little more than thirteen billion years after the creation of that first galaxy, we find ourselves roughly in the Jewish calendar year 5780. In AD 1656 the head of the Church of Ireland, Archbishop James Ussher, calculated that the world began exactly at a very specific moment: October 22, 4004 BC, at 6 o'clock in the evening. One wonders how the clock rang out 6:00 p.m. and what had just caused it to be evening, yet there you have it, the moment when time began. Of course, Ussher was using origin myths, believing that the earth and the entire universe came about as a splendid creation by God.

The modern story of the age of the earth starts with the seventeenth-century Danish Catholic bishop and geologist Nicolas Steno, who discovered that geological strata contained fossils. Steno certainly knew about Ussher's declaration, but his findings led Robert Hooke to suggest that the fossil record goes far back beyond the pyramids.

A hundred years later canals were being dug throughout Europe in accommodation of the Industrial Revolution. Canals had to be dug far deeper into the ground and bedrock than normal buildings of that time. The old geological picture changed when diggers exposed successive layers of bedrock showing signs of tilt and erosion. It gave geologists such as the Scottish natural philosopher James Hutton a new vision, suggesting that because heat is the prime agent of nature and the earth had been cooling for some time, the earth must be far older than scientists had previously thought. In fact, Hutton believed that there was no telling how old the world really was. He developed the idea known as *uniformitarianism,* the doctrine stating that all the geological processes that happen now are the same as those that happened any indefinite number of years in the past. "The present is the key to the past" were the British polymath poet and theologian William Whewell's words.[3] By that, Whewell meant that any past geological drivers, such as earthquakes, volcanic eruptions, and flood erosions, are the same geological drivers of the present and future. Their numbers and intensities never vary; at least, they roughly never vary. This idea refuted the early theories that species evolve by gradual change, as proposed in the nineteenth century by the English and French scientists Humphry Davy and Jean Baptiste Lamarck.[4] Whewell labeled the two opposing schools "uniformitarians" and "catastrophists." The uniformitarian principle guides rules for understanding geological formations as well as the rules for estimating the time it took for one of those formations to materialize.[5] This is how the Scottish mathematician and geologist John Playfair, a good friend of Hutton, put it: "Amid all the revolutions of the globe, the economy of nature has been uniform, and her laws are the only things that have resisted the general movement. The rivers and rocks, the seas and the continents, have been changed in all their parts; but the laws which direct those changes, and the rules to which they are subject, have remained invariably the same."[6] Playfair was best known in mathematics for what is called Playfair's axiom, a substitute for Euclid's parallel axiom—*in a plane, given a line and a point not on it, at most one line parallel to the given line can be drown through the point.* In the field of geology, however, Playfair is known for his 1802 publication of *Illustrations of the Huttonian Theory of the Earth,* a 528-page book that contains no pictorial illustrations but rather cases in support of, and cases of objection to, Hutton's point of view on the earth's age. In one such case Playfair makes an assumption, believed for ages before the twentieth century, that mineral substances are somehow constant.

The Scottish geologist Charles Lyell collaborated with Playfair. By examining the earth's planetary motion, Playfair and Lyell could not find evidence to support a beginning or an end of the earth.[7] Measurements were speculatively based on historical intuition that had almost nothing to do with science. Geological stratification was crude, but with hinting signals of an order of significant events in time. It was a time when scientists were beginning to ponder the thought that man's history on earth was attuned by the individual lifespans of fossil groups. Hutton never mentioned the possibility that fossil groups might give clues of significant geological events, perhaps because he believed that nature repeats itself in time cycles without admitting anything new. The twentieth-century paleontologist and evolutionary biologist Stephen Jay Gould wrote, "Fossils, to Hutton, are immanent properties of time's cycle."[8] Hutton's theories were based on the infant field of geology, where time was not simply the conceptual clock of relatively short historical changes but rather what the American writer John McPhee once called "deep time." McPhee coined the term in his book *Basin and Range* as the unfathomably long expanses of geological changes in rock formation. Referring to Hutton's uniformitarianism, by which geological shifts were incredibly slow, McPhee wrote that rock-earth changes required expanses of time "in quantities no mind had yet conceived."[9] Countering the older theory of catastrophism, uniformitarianism awakened scientists to new thoughts about the age of the earth.

Lyell's groundbreaking *Principles of Geology*, published in 1830, was a pictorially illustrated opposition to prevailing geological and theological views of how the earth was formed. Lyell believed that the earth's surface had formed through a huge number of small changes over vast periods of time, not through such large catastrophic changes as meteoric activity or land-devastating floods. Charles Darwin read Lyell's book while on his *Beagle* voyage and was inspired not only by the geological sense but also by, perhaps especially by, Lyell's belief that geological changes take place over exceptionally long time intervals. So, every past geological event was caused by the same phenomena that now cause geological events. Old events continue with the same frequency and intensity as always. The earth might seem different now in looks, vegetation, animal population, coastlines, and terrain, but, ignoring those slow changes in appearance, it has always worked in the same way that it works now.

Many cultures claimed starting times for time. For the Mayans, it was August 13, 3114 BC, as translated in Gregorian calendar dating. In the seventeenth century, the impression was that the universe is infinite in size and therefore time must also be infinite. Isaac Newton certainly believed that the universe is infinite. Lacking any historical records dating back further than the Trojan War or the great pyramids, many intelligent thinkers of Newton's generation believed that—if not the universe—then at least the earth must have been formed recently.

It's difficult to know Newton's religious belief; some biographers say that he was a heretic and that he surely did not believe that God recently created the planet we live on.[10] To him, the earth was formed by an accident of gravity and it just happened to fall into a particular spin and orbit giving us, roughly, a 365-day year.

Richard Bentley, a seventeenth-century classical and theological studies scholar with fervent scientific interests, thought that if universal gravitation is true, then all the stars in the universe would eventually be drawn to one another to form a single mass: any two stars close enough to each other would join into a larger mass, and this new mass would then pull other stars into an even larger mass. The process would continue till eventually every object in the universe would collapse into one single mass and presumably, thereby, rebirth the universe. The whole scenario would be, according to Bentley, divine intervention.

Newton had thought of this scenario. His inclination was to reconcile his theories of the universe with the Bible. His reply to Bentley set to rest any thoughts of a finite universe—and finite time. In a letter to Bentley dated December 10, 1693, Newton wrote:

> As to your first Query, it seems to me that if the matter of our Sun & Planets & all the matter in the Vniverse was eavenly scattered throughout all the heavens, & every particle had an innate gravity towards all the rest & the whole space throughout which this matter was scattered, was but finite: the matter on the outside of this space would by its gravity tend towards all the matter on the inside & by consequence fall down to the middle of the whole space & there compose one great spherical mass But if the matter was eavenly diffused through an infinite space, it would never convene into one mass but some of it convene into one mass & some into another so as to make an infinite number of great masses scattered at great distances from one another throughout all that infinite space.[11]

In his *Principia,* Newton computed that it would take a red-hot iron equal to the size of the earth fifty thousand years to cool to average earth temperature.[12] He was surmising this estimate from observations collected during the recent appearance of Halley's comet. So Newton explained to Bentley that if space is infinite and its matter were scattered evenly, mass would be pulled in all directions and not favoring any particular mass, no matter how big. He had not thought of the possibility of an expanding universe (a twentieth-century revelation), which would reinforce his argument. Is time infinite toward the past and the future? Or was there a real beginning of time? If so, what does that mean for existence? How did the universe then come to be from nothing, if God did not create it? And if there is a God, he must have already had time before creation, presenting a rather serious paradox of time.

Following the logic of Newton's arguments, it seems likely that time is infinite. But let's bring things back to earth. Bishop Ussher's calculated birthdate of the world as 4004 BC was really about earth, for in his time there was no serious distinction between the age of the universe and the age of the earth. By the last century, the earth's age was believed to be no more than 100 million years old. Lord Kelvin calculated that an earth-size mass, as molten as it was as a ball of fire to begin with, would have cooled to its present temperature in no more than 100 million years. His calculations, the best that he could do using what was then known about geology and thermodynamics, were off by a factor of more than forty-five. The age of the earth, as known today, is 4.543 billion years. Quite a difference!

We should relish the knowledge that our planet is that old. Had we lived in the nineteenth century with the feeling that the earth was so young as to be just 100 million years old, we would have had more of a paranoia that anything could happen to it. An asteroid could hit it at any time. With the world being so much older, we can feel that it has survived for such an extraordinarily long time without too many overwhelming consequences that the chances of surviving further are quite certain, if we treat it wisely, and don't blow it up ourselves.

Isaac Newton felt that true mathematical time was the mover of everything from stars to humans. To him it was some continuously flowing, enigmatic, forward-direction driver, an invisible river that propels everything to happen in lockstep with the myriad actions and events of the universe. He believed that, somehow, we can measure that mathematical time through observations of motion, as if the mysterious driver could never be known but its consequences could be measured. We still use the word *time* in the singular when thinking about Newton's baffling mover, even though we now know that it is not absolute and that its measurement submits to the phenomenon of time dilations that depend on relative speeds.

We know that the world turns in a measure of time and that all events and motions of the universe are orchestrated by time. Anything that changes does so by definition of time and, in particular, by a measurable duration.

Duration is tricky though. For Newton, it had one advantageous property: independence from the observer. By Newton's rules, two objects leaving the same place at time *A*, wandering in different directions and returning to that same place at time *B*, will have each traveled for the same amount of time, which should turn out to be *B* – *A* time units. That was a reasonable rule until the dawn of the twentieth century, when time changed from being absolute to being relative. Newton's rule about duration was no longer valid after Albert Einstein's theory of special relativity. For Einstein, objects leaving the same place at time *A*, wandering in different directions, to different locations, and returning to the same place at time *B* might not have each traveled for *B* – *A* units of time. It's a hard notion to absorb completely, given how much influence our language of time has on our thoughts.

The seventeenth-century philosopher and mathematician Gottfried Wilhelm Leibniz felt that time depends on events and relations between events with a tight bonding of space and time. Not anything like Newton's absolute. So we are left with an understanding that there are indeed two kinds of time, the subjective perceptual kind that gives us a reasonably clear grasp of the "now" and the objective conceptual time that gives us a sense of past, present, and future and their relationships.

The problem, as Leibniz saw it, was that if God created the material universe, he was selecting a time in which to do it and inventing a tool to coordinate all the events that would happen in that universe. Therefore, there must have been a time before creation that smoothly continued into the time after creation. "God does nothing without a reason," Leibniz wrote, "and since there is no reason assignable why He did not create the world sooner, it will follow either that He created nothing at all, or that He produced the world before any assignable time, which is to say that the world is eternal."[13] Leibniz was following an age-old question repeated by Saint Augustine: "What was God doing before he made heaven and earth?" Augustine quips about some of the answers, such as, "He was preparing Hell for people who pry into mysteries." But then, after a few humble excuses for not really knowing the answer, he gives us his thoughts. "What time could there have been that was not created by you [the Creator]? How could time elapse if it never was? . . . But if there was no time before heaven and earth were created, how can anyone ask what you were doing 'then'? If there was no time, there was no 'then.'"[14]

It seems that our respected philosophers of an earlier era believed that time began as if from a stopwatch that began to count only when its start button was pressed and that we are now living in the running of that watch. With that thought comes a perplexing feeling that whatever it is, that thing we are calling time is governed by a clock. As the clock moves, so does time. But time might be running out.

At 4.6 billion years, the earth is surely old, perhaps so old that it sees its time limited. For the past 150 years soapbox speakers around the globe have been howling, "The end is nigh." They see the end of time as the end of earth. There is some evidence that their prophecies are potential. There are signs of approaching environmental deterioration, and signs that the human existence the way we know it on this planet will soon play itself out. Time for humans on this planet seems to be limited. It might even come to an end. In October 2018 the UN Intergovernmental Panel on Climate Change reported that more than one-tenth of the world population will be persistently malnourished by climate-related shocks and disasters.[15] Of course there have always been wars, droughts, monsoons, tsunamis, and famines, but current manifestations of a changing climate paint a very different future for human existence. "This long-term view shows that the next few decades offer a brief window of opportunity to minimize large-scale and potentially catastrophic climate change that will extend longer than the entire history of human civilization thus far."[16] Without international policy for climate change or spectacularly clever technological advances to alter the climate, the future of humankind might be darkened for many thousands of years. It's not just rising sea levels that will cause significant impact on life, but ocean temperatures and increased acidity limiting the ability to grow certain mainstay crops such as rice and wheat. For the past two hundred years we have been tampering with the natural cycle of climate that has kept the earth at stable carbon dioxide levels for hundreds of thousands of years. Current carbon dioxide levels are now 69 percent higher than they were 252 million years ago at the time of "the most severe biodiversity crisis in earth history," the Permian extinction, which wiped out more than 70 percent of the earth's vertebrate species and destroyed all but 4 percent of marine species. Wildfires and methane gas spread across the entire globe for more than 200,000 years. With 83 percent of all genera extinct, recovery took another 2 to 10 million years.[17] We are now heading for a time when we will be losing landmasses along coastlines where fertile soil will be polluted by salt leaching from encroaching seas. Every year from now to an indefinite time we will be losing inhabitable earth. Time for making significant changes is running out. Hurricanes will grow stronger; droughts will spread wider and inland. It will not be the total end of time on earth, but it may be the end of time as we know it. Where will 1.5 billion people go in the next 50 years when temperature rises to uninhabitable levels, floods take over habitats established for hundreds of years, drinkable water becomes scarce, and resulting mass migration causes uncontrolled wars all over the globe? What will time be like for those poor folks whose lifespans will be shrunken to the average age of a horse? When things get worse, were will they go? Stephen Hawking tells us that we'd better look for another planet. Speaking in Trondheim, Norway, at the Starmus Festival IV, nine months before his death on March 14, 2018, he warned, "If humanity is to continue for another million years, our future lies in boldly going where no one else has gone before. . . . Spreading out may be the only thing that saves us from ourselves. I am convinced that humans have to leave Earth."[18] If we do ever leave—and make no mistake about it, sooner or later we will leave—we will have an enormously difficult problem of changing everything we ever knew about time, to say nothing about our adjustments to gravity, oxygen, and whatever else our exile will have in store.

# INTERLUDE: PRISONERS IN TEXAS AND OKLAHOMA

Prisoners do have time, lots of it, concrete time, and they focus on it. In one episode of the Netflix series *Orange Is the New Black,* we find ourselves looking in on a fictional prison, Litchfield, where an inmate, Lolly, who hears voices, builds a cardboard refuge she calls her time machine. Healy, the prison counselor, finds her, sits beside her in the box, and softly tells her that everyone wants to go back in time, but it's not possible. For many of the prisoners I interviewed, there are regrets and wishes for rewrites of the past. They all wanted to go back to the moment when things could be made different, before they made their mistakes. But time points toward the future, not the other way around. The present is a peculiar instant, the single moving step on up-escalators through life. It moves between levels of age, as if the future is in front, and the past everything behind.

Not too long ago, I was a guest of a monthly interfaith service at the Southern State Correctional Facility in Springfield, Vermont, where I learned that some inmates have exceptionally acquired senses of what divides the past from the future. Of the 353 inmates, 5 came to the service. They were dedicated to moments of self-reflections of their life-changing mistakes. One had served sixteen years. I didn't ask how he thought of time during his long internment, nor did I ask how he thought of the time he had left to serve, another twenty-four years. But that person indirectly revealed a thought that caught me by surprise.

"Strange for me to say," he volunteered. "My being in here is a good thing. I don't know what harm I might have done, if I were not."

Another inmate agreed. "I was on serious stuff," he said. "I thank the Lord I'm in here."

Two others had similar thoughts; the fifth, annoyed with the system, claimed innocence. The service, which lasted ninety minutes in a chapel that looked like a utility room, was more akin to a confessional than a prayer meeting. I had never before been inside a prison, so, as in the movies I had seen, I expected to come face to face with hardened criminals, half giving excuses for their crimes, the other half claiming innocence. From the moment the five entered the chapel, though, as time went forward, a seed of sympathy grew into overwhelming compassion. I was surprised at how I first saw them as numbered convicts and later as humans with names. I'm not a teary person, but their demonstration of personal regrets, their testimonies of mental what-ifs, their fantasies of backward time travel to take those paths that were not taken, left me with wet eyes.

I spoke with several prisoners in Texas and Oklahoma penitentiaries who had served sentences for extraordinarily long periods. Their views were similar.

# 6

## GUTENBERG'S TYPE (TIME IN THE FIRST INFORMATION AGE)

Thomas Aquinas, the Italian Dominican friar, theologian, philosopher, and most influential sage of medieval times, believed that God is the cause of time. For Aquinas, reason and faith are both gifts from God that are proofs of his existence. That belief was the official papal decree, designed to suppress all contradictions to church teaching. It proclaimed that Aristotle and the Arabs were infidels and that God is the maker of motion and, hence, of time. Paganism had been a crime in the Roman Empire, ever since 391, when the emperor Theodosius I issued an edict declaring paganism to be "a crime of high treason against the state, which can be expiated only by the death of the guilty."[1] In Alexandria one year later, Christian marauders torched the library of the pagan temple of Serapis (the Serapeum), burned more than three hundred thousand scrolls, and murdered in the streets several of the museum's scholars, including the celebrated Hypatia, the first female mathematician identified in the historical record.[2]

As the Roman Empire broke apart, the Islamic world rose. Muslims conquered the south of the Mediterranean from Syria and Mesopotamia to Spain, expanding well beyond the limits of Roman civilization, spreading into Asia and Africa. Arabs brought inventions back from China and India, advanced astronomy, introduced the Hindu notion of zero, invented algebra, developed the chemistry of metallurgy, and designed the mizzenmast to speed their ships. On November 27, 1095, however, Pope Urban II addressed a large crowd in a wheat field in Clermont, France.

"Jerusalem is the navel of the world," he called out, "A land which is more fruitful than any other, a land which is like another paradise of delights. This is the land which the Redeemer of mankind illuminated by his coming, adorned by his life, consecrated by his passion, redeemed by his death and sealed by his burial." And with a passionate plea, Urban II incited the crowd to take up arms against all heathens. "This royal city, situated in the middle of the world," he continued, "is now held captive by his enemies and is made a servant, by those who know not God, for the ceremonies of the heathen. It looks and hopes for freedom; it begs unceasingly that you will come to its aid. It looks for help from you, especially, because God has bestowed glory in arms upon you more than on any other nation. Undertake this journey, therefore, for the remission of your sins, with the assurance of 'glory which cannot fade' in the kingdom of heaven."[3] The Crusades brought magnificent spoils of war to churches and monasteries all over Europe. Among the treasures were silks, perfumes, spices, and books written in Arabic—translations and transcriptions of Greek and Egyptian scrolls stolen from Arabian libraries and, in particular, Aristotle's eight books on physics. Only theologians were openly permitted access to Aristotle's pagan works, which were banned in most of the new universities of Europe. Heresy was considered a serious crime, and the punishment for reading Aristotle was imprisonment for life.

That changed in 1262, when Thomas Aquinas was in residency at the papal court in Orvieto, near Rome. There he met the Flemish Dominican William of Moerbeke, who had translated several works of Aristotle from Greek to Latin. Although Aristotle was a pagan, Aquinas found a way to write commentaries and brilliant explanations of what Aristotle had in mind. By the early fourteenth century Aristotle's works were permitted, even fashionable. Aquinas's commentaries and interpretations had made them acceptable to the church since they did not really interfere with church teachings. For a while, physics was fully influenced by Aristotle's *Physics*, describing motion as conditioned by time, comparing "velocities" as one being quicker than another.

Johannes Gutenberg's movable-type printing press changed the course of intellectual and scientific information. It gave Europe its first information age. In 1436 replaceable wooden or metal letters and the invention of plant-fiber paper, brought to the West from China via the Silk Road, were responsible for the publication of more books in the sixteenth century than had been produced in the thirty-five-hundred-year period since the first Babylonian author produced the earliest cuneiform tablet.

Teaching at that time was dictatorial, and rote memorization of Aristotle's works played a central part in the curriculum. The seven liberal arts—grammar, logic, rhetoric, arithmetic, geometry, music, and astronomy—were required, though how much of each was a matter under local control. This rote learning numbed the intellect so severely that nobody thought to criticize the classic works of science, especially the unshakable doctrines of Aristotle. Moreover, except for rote learning of arithmetic and computation, mathematics was completely neglected. "The names of Euclid and Archimedes were empty sounds to the mass of students who daily thronged the academic halls of Bologna, the ancient and the free, of Pisa, and even the learned Padua."[4] The Italian humanists, who studied the principal literature of antiquity for literary content—as opposed to theological matter—accepted printing with scorn: "Printed books seemed a cheap substitute for their beloved manuscripts, nor did they wish any enlargement of the reading public to include persons without taste. Taste, style, manner, correctitude, *aplomb* were set above more substantial attainments."[5] But the works of Archimedes, which had been copied into Greek in the ninth century and translated into Latin in the fifteenth, were now being printed and sold throughout Europe. These works were beginning to inspire a new generation of independent thinkers to rethink old doctrines of motion, mathematics, and time.

---

Science in the late sixteenth century was still rooted in Aristotelian creed. Mistrusting new ideas, scholars continued to believe that all worthwhile knowledge was already expressed in library documents written by the ancient sages.

Then came Galileo Galilei. By measuring the weight of water leaked from a tank during the period an object fell, he debunked the existing theory of his era that heavier objects fall faster than light ones and in its place instilled a counterintuitive truth that all objects fall at the same acceleration and speed. The Australian mathematician Robyn Arianrhod tells us, "Others had wondered whether gravity had the same effect on the motion of all objects, but Galileo's result was compelling in its precision. It is experimental and mathematical precision that differentiates physics from all other forms of enquiry into the nature of physical reality."[6]

As one celebrated legend has it, on Sunday mornings Galileo would leave his house near the Porta Florentina, walk along the cobblestone quay of Santa Maria della Spina on the banks of the Arno, cross Filippo Brunelleschi's magnificent Ponte a Mare, and attend Mass at Pisa Cathedral. The cathedral is architecturally plain, but its nave is large and ornate, with a three-tier, thirty-candle bronze chandelier. Before each Mass the giant fixture would be lowered by a chain through its center for candle lighting, and then raised. For a considerable number of minutes, perhaps five or even ten, the chandelier would gently sway, hardly noticed by worshippers. But Galileo did notice—at least he did on one Sunday morning in 1583, when he was sitting through a dull sermon with his wandering thoughts and watching the chandelier swing. The nineteen-year-old marveled at its motion, timing the oscillations against the timing of his pulse. This was a revelation about how to measure time. As the fixture slowed, the duration of each swing remained constant. Shorter swings were simply slower. He correctly guessed that the time required for any complete swing of the chandelier depended solely on length. The story must be fictional, for the cathedral chandelier was not installed until 1588, five years after Galileo claimed the discovery. Vincenzo Viviani, a pupil and friend of Galileo, perpetuated the story, saying, "Having observed the unerring regularity of the oscillations of this lamp and of other swinging bodies, the idea occurred to him that an instrument might be constructed on this principle, which should mark with accuracy the rate and variation of the pulse."[7] And from then on, the myth continued, suggesting that Galileo used his pulse in his experiments with falling objects. True, clocks with an accuracy that was needed to measure the speed and acceleration of falling objects were not available. But Galileo was Galileo, so he found other ways to measure time.

It was an era of mixed contradictions. The century before was filled with new explorations. The Americas were discovered and conquistadors found fruits, vegetables, and nuts never seen in the Old World. Europeans were given a new view of the world. Vasco da Gama sailed around the Cape of Good Hope. Ferdinand Magellan sailed clear around the entire world.

The vast Pacific was discovered. We in our twenty-first century cannot put ourselves in the place of their thoughts, for the stretch of knowledge from theirs to ours is vast. We can go from New York to Beijing in less than fourteen hours, so how can we even think that stones on the other side of the earth would not fall to the ground. A hundred years from now people will read that a trip to the other side of the world took fourteen hours and chuckle at how unbearably slowly we traveled. The pace of time accelerates with technological momenta. But back then, in Galileo's era, the big ideas of the world came from scholars who sat in dimly lit university libraries and secluded monasteries questioning the laws of nature and how the continent of America could exist without being mentioned in the Bible or in the writings of Ptolemy and Aristotle. Hesitant to defy their Bible, they challenged the wisdom of established classical intellectual teachings and began to investigate nature by direct observation.[8]

Perhaps that is why young Galileo did not accept assertions without examining, weighing, and reasoning their truths. He despised university training, which professed truth by authority and regarded any contradiction to Aristotle as blasphemy. His teachers found him obstinate and uncooperative. Secretly, on his own, he read the first six books of Euclid and continued to study mathematics with great passion, believing it to be the best means to understand nature's secrets.

Soon after, with the title professor of mathematics, he felt the concept of time and motion to be central to the scientific understanding of almost all natural phenomena. He read a treatise by the Venetian mathematician Giovanni Battista Benedetti called *Parisian Physics* that explored a new notion of the physics of motion: air was not the cause of motion; rather, it was the object itself that contained the cause. Benedetti's book rekindled speculations on mathematics and physics established two centuries earlier by the influential French philosopher Jean Buridan at the University of Paris. That book, along with the inspirational powers of reading Euclid and Archimedes, encouraged Galileo to replace Aristotle's empirical methods and ideas on time and motion with observational methods and mathematical reasoning. "The method that we shall follow," Galileo wrote in his *De Motu* (On motion), "will be always to make what is said depend on what was said before, and, if possible, never to assume as true that which requires proof. My teachers of mathematics taught me this method."

Aristotle claimed that two bodies made from the same material would fall at speeds that are proportional to their sizes; hence, a large piece of gold should fall faster than a small piece. By Galileo's logic such a notion is ridiculous. "How ridiculous this view is, is clearer than daylight," wrote Galileo, who then argued that if two bodies of the same material and weight were let go in a medium, then Aristotle would be forced to say that the two bodies together would descend faster than either one alone. "What clearer proof do we need of the error of Aristotle's opinion? And who, I ask, will not recognize the truth at once, if he looks at the matter simply and naturally?" His reasoning was presented so simply and naturally that we must wonder how Aristotle could have missed such a clear argument. Surely, if one object is a thousand times heavier than another, the velocity of one is not going to be a thousand times faster than the other. Through proofs and logical reasoning, Galileo destroyed Aristotle's principles of physics. In his *De Motu,* he asserted that Aristotle's assumptions were false and that not only was Aristotle wrong about the speeds of falling objects but he was wrong about almost everything to do with locomotion. Eventually, Galileo's thinking moved from "how" things moved to "what" moved them.[9]

The Aristotelian doctrines of motion began to crumble as more and more scientists and natural philosophers were basing judgments on real world experiments rather than purely intellectual reasoning. Inconsistencies popped up with each new experiment. Each new inconsistency was met with a tailoring of meaning. "Well, Aristotle meant to say . . . ," his supporters would say, until a blitz of discrepancies forced too many unnatural alterations into a quilt of conflicting patches of truth.

One way to define time is by defining its relation with space. Galileo did that in his *Two New Sciences* in Third Day, Theorem I, Proposition I, when he wrote: "If a moving particle, carried uniformly at a constant speed, traverses two distances the time-intervals required are to each other in the ratio of these distances."[10]

At the beginning of Third Day, he tells us:

> I have discovered by experiment some properties of [motion] which are worth knowing and which have not hitherto been either observed or demonstrated. Some superficial observations have

> been made, as, for instance, that the free motion of a heavy falling body is continuously accelerated; but to just what extent this acceleration occurs has not yet been announced; for so far as I know, no one has yet pointed out that the distances traveled, during equal intervals of time, by a body falling from rest, stand to one another in the same ratio as the odd numbers beginning with unity.[11]

Clearly, from his own writing, he did not know that his proposition had been established almost three hundred years earlier by a group of four young mathematicians at Merton College, Oxford. That was a time when gunpowder, firearms, and cannons were appearing in Europe to close out the era of armored knights in fortified castles. It was a time of great interest in the science of trajectory motion and targets. The first cannon was probably fired at just about the time that the four mathematicians were sharing their ideas on the mechanics of motion. It might seem surprising that the physics of trajectory motion for archery was not truly understood, other than by exposure to enough experience, after a multimillennium history of hunting with weapons from spear to slingshot to crossbow. But cannon fire was different. Every firing was expensive—and deadly. Devastating collateral damage called for better accuracy. Thomas Bradwardine, William Heytesbury, Richard Swineshead, and John Dumbleton, known as the Merton College four, sharing their ideas on the mechanics of motion, pointed out that the distances traveled, during equal intervals of time, by a body falling from rest with speed increasing at a constant rate, stand to one another in the same ratio as the odd numbers beginning with unity. In each and every increment of time, the object picks up an equal increment of velocity. In other words, the object moves twice as fast in the second second as in the first, three times as fast after the third second as in the first, and so on. It was still an era when science was most convincing through rational arguments. The causes and effects of motion were beginning to be distinguished and rationally understood; the ideas of instantaneous velocity and uniformly accelerated motion, and the abstraction of time, were emerging for what was to become (three hundred years later) one of the prime motivations of calculus. And then came Isaac Newton.

## INTERLUDE: TIME LOCKED INTO THE PRESENT

I met Clint Barnum, a seventy-year-old ex-con, in a café one afternoon just after the third nor'easter of March 2018 was dying down. A stonemason on the outside, skilled at building attractive stone walls, Clint had finished serving seven years of his seven-to-ten-year plea deal sentence for voluntary manslaughter. Of the many ex-cons whom I interviewed, I found Clint to be the most reflective of his past experiences with time. Although he was never in solitary, his thoughts of time in prison were buried in the routine of his job as a janitor. Time for him inside prison went at just the same speed as it did outside.

"There is nighttime, my time, and daffodil time," Clint wisely professed. "Are you talking about a plant or a human? What do you mean by time?"

"What was time like for you?" I asked.

"The time in there is so locked into the present. You think of the past but know you are in the present all the time. In prison you know exactly what's what. Inside prison there is not much change. In getting out, there is so much change. Each day takes a week. Cars go by. You can touch trees. Each day is not like the one before."

For Clint, nothing was stable in prison. He spent his time in different cells, but his routine job filled his moments, even though it was the same job over and over again. So time for him went by at an expected pace, not fast, not slow.

"In prison there are all sorts of people," he told me. "Some have done some really bad things, some have repented, some are repairable, some struggle back and forth with the background from which they came, some have been raped by their fathers or grandfathers, and some are so broken they are not repairable. You don't hear about those *non-repairables* so much because they are mostly in solitary."

"Yes, but I'm interested in time, how prisoners might think about time, the times they have left to serve, the future, the past, and how fast or slow they feel time moving."

"They are in for all kinds of reasons," he insightfully explained. "So when you talk with them, you will get all kinds of takes on this or that, and especially on time. There's a whole spectrum of views when you get to the prison population, which is simply a microcosm of the outside world. How will you get any kind of consensus of the human understanding of time when you talk with prisoners who see the world so differently when they are inside and when they are out?

"Whoever you are in prison, you fill your world with what you see and what you learn from the internal world. You know there is some external world that faded with time. Time in jail has something to do with age. Younger inmates are off the wall, wilder. Older are more mellow. They learn more who you are and have a sense of honor. Younger guys don't have internal anchors, but with age they see the patterns of life. Souls mature with age, and with that maturing you understand what you have to achieve in your lifetime. You'll get six very different stories from six different inmates. There is a huge variety of understandings and experiences of . . . "

"Yes, I am getting six different stories, Clint," I interrupted. "But I'm interested in yours. What is time for you?"

"Time?" After a long, reflective pause, he said, "Time is linked in some way to consciousness. It is the measure of life. We are conscious beings placed in a physical world that has no consciousness, a world of rocks and plants that change in connection with conscious beings who are trying to make sense of the connection. If you take away consciousness, you are left with nothing but change. There is an abyss between pure consciousness and pure mindlessness, a disconnect."

# 7

## ENTER NEWTON (ABSOLUTE TIME)

For so many centuries before Newton, villages, towns, and cities throughout the world had measured time as divisions of the daylight and darkness. The first step in measuring time was to start with the definition of day, month, and year. So, in the past, one could simply take the span from sunrise to sunset and divide by twelve to get something that we might call the hour and then do the same with the span of night. But that hour was different from ours. It differed from daytime to nighttime, from day to day, from place to place, from season to season and, by its definition, depended on the sun, the earth, and the stars. The sun appears to move across the sky at a uniform speed without jumping or speeding up. The same seems to happen with the stars. They give the impression that they are fixed with regard to one another, yet as a whole they look as if they move uniformly across the sky.

It was an effective way of keeping time for simple things like waking, sleeping, resting, working, eating, and, let's not forget, praying. But dividing the span of daylight time by twelve meant that there had first to be some numerical measure of the timespan of daylight time. How else could anyone divide something by twelve when the "something" is not a number? The only way to do that is to connect time with the path of the sun or the motion of the stars. Daylight hours are relatively easy to measure by a sundial. Mark the trajectory of an obelisk's shadow from sunrise to sunset. Now you have a length. Divide the trajectory into twelve parts. Each of those parts represents what could be called an hour. Nighttime hours are trickier; the trajectory of a star must first be mapped, marked, and divided. For most people, nighttime hours had little importance. The night was the night, dark, with little need for keeping a schedule. You slept when you got sleepy or bored.

To create his theory of gravitation and dynamics, Newton needed a functional definition of velocity, which meant that he had to introduce both absolute space and absolute time. For him, there were no alternatives, even though he certainly knew that any observable body moves from one place to another by relative change. His absolute space was "some motionless thing" distinct from physical bodies. Very early in the *scholium* of his *Principia* (published in 1687, a half-century after Galileo's publication of his *Two New Sciences*), he advanced his distinctions of time, conceding that there is such a thing as relative time and the apparent time that we live by, though holding that they are different from what he called absolute mathematical time, a time that by "its own nature flows equably without regard to anything external."[1] He believed that the words *time, space,* and *place* were too often used through their relations with how they affect the senses, and he was concerned that the senses tend to build confounding missuses.

Astronomers saw time differently. They simply and sensibly measured time from dawn to dawn and divided by twenty-four. They knew that the sun (moving around the earth as they imagined the mechanism of the solar system) would have traveled 360 degrees from dawn to dawn. That defines the hour at exactly 15 degrees of the sun's fancied circle around the earth. The simplest instrument to measure 15 degrees of the sun's arc is the hand. When someone holds a hand out to the sky at arm's length, bending all fingers except the pinky and index finger, those two extended fingers point to approximately one hour of the sun's movement across the sky. Of course, that is assuming that the sun travels around the earth in a circle. Circles, ellipses, . . . it doesn't matter. Appearance one way or the other is the same. The observer on earth does not have to know whether the earth is orbiting the sun or the sun the earth. A similar measurement can be made at night under the deceptive assumption that the stars are orbiting the earth at the rate of 15 degrees per hour. Keeping records of hours during the day is not so difficult, as long as we accept our measures as crudely approximate.

Newton's true time is different. True motion is independent of the observer, if there is such a thing in our perpetually expanding universe. Relative time is time measured by the apparent motion of the sun relative to the earth, or the stars relative to the moon—that is what our clocks measure. All clocks measure time that depends on the appearance of celestial events. Even the most sophisticated atomic clocks are calibrated to celestial motions that humans have observed and recorded. What if there is a system that could give us some sort of time parameter that is independent of observation, a system that recognizes the motions of all celestial activity without violating any laws of physics? Such a system would suggest that the time we currently know and live by is merely an illusion. It would imply that the speed of light is both a constant and independent of the observer. Thanks to late nineteenth- and early twentieth-century physics, we now know more about the speed of light and time's dependence on the observer to realize that that would be a contradiction, and so we must agree that there is no such system. Newton could not have known of that contradiction, but he did believe that there could be a time parameter independent of the observer: *true time*. Relative time, or any time measured by celestial motion, can be measured in several ways, but *true time*, if there ever was to be such a thing, is the unifying time that underlies all measures of time.[2]

Newton's notion of absolute time is easily confused with his notion of true time. He gave this to ponder: "Absolute, true and mathematical time, of itself, and from its own nature flows equably without regard to anything external, and by another name is called duration: relative, apparent and common time, is some sensible and external (whether accurate or unequable) measure of duration by the means of motion, which is commonly used instead of true time, such as an hour, a day, a month, a year."[3] Absolute time is a mathematical time approximated by celestial time, a time measured by, say, the solar day or the celestial sphere, a time used by physics and astronomy to predict future celestial events. As absolute as it is, it is still just an approximation fitting for temporal dealings of the solar system and relatively near stars, not the whole universe and everything that exists beyond, if there is a beyond. Only by approximation does absolute time apply to bigger pictures. True time is the time that applies to the entire universe, to any time frame anywhere.

Newton's absolute time was independent of events, an absolute that defines everything in our neighborhood of the universe and distinguished from the conception of time in life's environment. He needed absolute, mathematical time to work with accelerations of moving bodies. He had to distinguish between true and relative motion and hence between true and relative time. In his view, true motion of an object cannot be generated or changed except by a force operating on that object. That's not the case with relative motion. Relative motion can be generated or changed without the need of a force. By modern knowledge of physics, and strangely by Newton's own laws of physics, there are hidden forces that act to make relative motion seem possible without forces. What makes this more confusing is that just several pages into his *scholium* is a strange example of what he calls relative motion. Strange, but brilliantly thought-provoking. Imagine a pail, half filled with water, hanging from a cord. The pail is turned to twist the cord tightly. When let go, it will spin to untwist the cord. At first the surface of water will be flat. As the spinning accelerates, the surface becomes more and more concave, and the rotation of the water stays with the rotational motion of the pail, even after the speed of spinning subsides. In Newton's words, the pail was "gradually communicating its motion to the water."[4]

It is a peculiar use of words, a misguided notion that there are no forces on the water to make that surface change from flatness to concaveness. His point is that before and slightly after the spinning occurs, there is relative motion between the water and the pail, meaning that relative to the water, the pail is spinning. As the spinning continues the water begins to spin with the pail. When that happens, relative to the water, the pail is stationary. Newton thought that the relative motion of the pail had no effect on the surface of the water, but during the spin, when there was no relative motion between the pail and water, the water formed a concave surface. That was his proof; relative motion was changed without a force. By that example, he concluded that rotational motion is absolute, simply because during rotation, from the point of view of the water, the pail is at rest, and in "performing its revolutions in the same times with the vessel, it becomes relatively at rest in it."[5]

Newton knew nothing about surface tension and friction between water molecules and pail molecules, so he assumed that the pail could somehow communicate its motion to the water. By this "communication" he concluded that rotational motion was absolute in the sense that if the one body was rotating, then the other had to keep up its rotation in order to adhere to an absolute motion, as if nature somehow abhorred the possibility of a relative motion. From this, and the thought that time and motion were dependently linked, he concluded that time must also be absolute. The nineteenth-century physicist Ernst Mach, who certainly knew about surface tension, disagreed with Newton. In *The Science of Mechanics,* Mach wrote that Newton's rotating pail experiment was simply a specific model involving a pail with sides of too little mass to have an influence on the water molecules. "The one experiment only lies before us," Mach wrote, "and our business is, to bring it into accord with the other facts known to us, and not with the arbitrary fictions of our imagination." A thin pail with minimal mass would have no noticeable forces influencing the water, no gravitational attraction. His comments referred to Newton's conclusion about absolute time and motion. It was a clever argument raising a broader thought: that the vessel and water were just two items in a very big universe, and so, relative to other objects in the universe that presumably also could be spinning, it might not be clear that the vessel was actually spinning. He concluded, "No one is competent to say how the experiment would turn out if the sides of the vessel increased in thickness and mass till they were ultimately several leagues thick." In other words, the sides could be the hidden variable playing a gravitational role.[6]

Mach also argued that both water and bucket have mass $m$. The velocity of the water starts off at $v_1$ and ends up at $v_2$. To get the water's velocity up to $v_2$, a force (no matter what that force is) must be applied, and that force is given as $p = m(v_1 - v_2)/t$. Time $t$ is included in the measuring of that force, and also, there is relativity between the two velocities. Hence, Mach says, "*All* masses and *all* velocities, and consequently *all* forces, are relative."[7]

Let's not forget that Mach's *Science of Mechanics* was published two hundred years after Newton's *Principia.* But Newton's contemporaries were also critical of Newton's absolute time.

Think about how Newton was able to corner mathematical models through discrete intervals that approach continuity. His derivatives and integrals are just that, packets of wispy discrete space and time ratios, intervals divided by durations, hints of what they would approach if those durations vanishingly approached no time at all. By that invention, continuous motion could be understood, at least mathematically. But the mathematics of calculus is simply a marvelous trick that makes it seem as if the continuities of time and motion are understood. It cannot break the illusion of continuous motion by giving it the status of discrete movements appearing to be continuous when durations are very, very small. Calculus gives us the practical way around the dilemma of what makes motion continuous. It justifies the procedural methods to handle continuity, but it does not give us a handle on what those small elements of time really are. We can let an interval of time $\Delta t$ approach zero, but can an interval of real time actually do that? The beauty of mathematics is that we can use it to think so, even though we don't know what time is, really.

---

Two centuries after the publication of Newton's *Principia,* the French mathematician and philosopher of science Henri Poincaré felt that all ways of measuring time should be respected as equivalent as long as time is defined to preserve both the conservation of energy and Newton's law of gravitation. At the end of the nineteenth century, Newton's laws were still considered experimental truths, and time was believed to be only approximate. Whatever we use as a means of measuring time should simply be the one that is most convenient.

> Time should be so defined that the equations of mechanics may be as simple as possible. In other words, there is not one way of measuring time more true than another; that which is generally adopted is only more convenient. Of two watches, we have no right to say that the one goes true, the other wrong; we can only say that it is advantageous to conform to the indications of the first.[8]

General relativity theory separated local time from cosmological time. A local clock directly measures local time. Cosmological time has no physical reality other than the relative motion witnessed by the observer in the vicinity of the measured motion. "It is the *coordinate time* in an arbitrarily chosen system of space-time coordinates."[9]

PART

# THE PHYSICS

There is no absolute space, and we only conceive of relative motion; and yet in most cases mechanical facts are enunciated as if there is an absolute space to which they can be referred.

There is no absolute time. When we say that two periods are equal, the statement has no meaning, and can only acquire a meaning by a convention.

Not only have we not direct intuition of the equality of two periods, but we have not even direct intuition of the simultaneity of two events occurring in two different places.

—*Henri Poincaré,* Science and Hypothesis

# 8

## WHAT IS A CLOCK? (TIME BEYOND THE OBSERVED)

I was twelve years old when my father, an armchair philosopher, relayed a myth that there were only three people in the world who understood Einstein's special theory of relativity. The myth was popularized by an unsubstantiated rumor that the renowned early twentieth-century English physicist Sir Arthur Eddington quipped, "Who's the third?" At twelve, I took my father's words to be sacred truth. I believed everything he said. I still cannot wear socks to bed because he once professed that feet need to breathe at night and that socks prevented feet from breathing. Perhaps that's why I started my first year of teaching a course on mathematical physics to undergraduates using material from Einstein's 1905 paper "Zur Elektrodynamik bewegter Körper" (On the electrodynamics of moving bodies).[1] It was in defiance of my father, and bravely naive for a newly appointed professor of mathematics. There I was, team teaching with a physicist colleague from pages of Einstein's famous paper translated into English, testing whether my undergraduates could increase the number to nine to include myself. In a few weeks our class of just five students was able to absorb almost the whole paper and understand it as easily as if it were analyzing the logic of Lewis Carroll's *Through the Looking-Glass.* There is nothing in it that an undergraduate who has been through a good course in advanced calculus cannot understand. Indeed, it is a beautifully written paper and a model for how papers in physics should be written, complete with an informational narrative between the mathematical complexities, hardly more than a few partial derivatives surrounding Maxwell's equations that must be understood.

It is possible that the myth was related more to Einstein's general relativity paper of 1915, a much harder to absorb idea and one that is a short bit beyond the prospective understanding of a typical undergraduate, but I don't think so. The myth might have been true before Einstein's papers were translated and edited into languages other than German in *The Principle of Relativity,* which appeared in print in 1923.[2]

But here is the problem, and possibly the seed of the myth my father was perpetuating. The math was clear, but the counterintuitive implications of its paradigm shift were beyond what any of my five students would accept. At issue was the wild thought that somehow space and time were not absolutes and that somehow every understanding of space and time that had jelled in their young minds had crystallized into beliefs that could not be shaken. *Your yardstick shrinks with speed? How absurd! Your watch slows? Well, then, measuring is meaningless! Science, whose raison d'être rests so much on the accuracy of measurement, must be hollow!* Perhaps that was what generated the great myth. It wasn't about understanding the argument; rather, it was about not accepting an argument in conflict with a strongly established intuition bias. My students had to accept a universe that seemed far too fantastical. Even though they knew that the Michelson-Morley experiment proved that space had no ether, that imaginary hidden medium in which light or gravitational forces might propagate, it was hard for them to imagine otherwise. The ether had a conceptual function by which one could envision motion in a medium. How can one picture the propagation of motion in something that is nothing? That was the confusing issue for my students. The math was fine. The image of propagation was the problem.

One other problem was the fact that our consciousness is really a three-dimensional cross-section of a four-dimensional world. Our pictures of reality are always fixed in a three-dimensional snapshot where time moves along to the next snapshot with regular, unwavering uniformity. We sometimes imagine distortions in that uniformity of time when life has its excitements and boredoms, but our experience with clocks and their precisions tells us that those imaginings are merely illusionary distortions of real time.

By the turn of the twentieth century all notions of an existing ether were gone, yet time was still viewed as a continuously flowing *something,* independent of any person, an absolute, a kind of universal clock that marked the trajectories of everything that ever moved, Newton's time. That view changed forever to become one where time was many different times, each depending on each thing that moved. Even the oddest notions to accept were beginning to be believed: that time stays still at the speed of light; that for the photon, time does not move at all; and that, unbelievably, a person traveling at near the speed of light would hardly grow old, relative to an earthbound stationary observer.

Shouldn't we come to some understanding of relative velocity in the same way we understand the speed of light (normally denoted as $c$)? Shouldn't it be natural to suggest that if someone—let's call him Captain Apollo—were on some spaceship moving at half the speed of light, his measurement of the speed of light would be half the speed of light—that is $c/2$? The problem is that Captain Apollo's measuring instruments are foreshortened just enough to keep the speed of light intact and constant. A somewhat counterintuitive notion. Captain Apollo's yardstick will have shrunken relative to what it was when his ship had not been moving. Moreover, his clocks will be running slower. We might expect him to measure the speed of light as $c/2$. If Captain Apollo were to measure the speed of light while traveling at a speed of $c/2$, he would find that that $c$ is still 186,300 miles per second. No matter what his speed—from his point of view, from ours, or from anyone else's—$c$ will always be measured as 186,300 miles per second. That marvel is the core mark of special relativity. Before Einstein, that notion was so foreign to all Western thought that it took a considerable amount of perceptive thinking even to begin to imagine such a possibility—that time does not enjoy the special privilege of being absolute. However, within the bounds of small speeds, time does pass with pretty strict precision.

Einstein's theory unlocked the gates surrounding that established intuition bias with keys that earlier relativity theorists didn't have. By 1900, Henri Poincaré had already worked out the principles of relative motion, predating Hermann Minkowski's space-time (described in chapter 10). Hendrik Lorentz, even with his working out the mathematical equations of length contraction, believed in the ether and believed as well that time dilation was simply a mathematical artifact and thus missed seeing those keys.

Space-time was more critical to general relativity than to special relativity, but Einstein's 1905 paper on special relativity implied a quandary over remote simultaneity, a paradox for clocks suggesting a model thought experiment. Take two clocks, *A* and *B,* that are identical in construction and mechanization. Both operate in exactly the same way. Both *A* and *B* start off side by side. Suppose that *B* moves at a constant speed *V* along a curve that wanders, say, for a mile and returns to rest alongside *A*. Clock *A* in its reference frame will record *B*'s journey as an interval of $t$ seconds. Clock *B* in its reference frame will record its journey as $t'$ seconds. But according to Clock *B,* $t'$ will be smaller than $t$. If the velocity *V* is large, $t'$ will be *significantly* smaller than $t$. Even with this understanding, we can imagine how deeply this idea cuts into a young person's intuition bias.

Yet still, human perceived time is locked in that living feeling of absolute time, time that ticks away here and there as far as the farthest stars, independent of velocity, place, temperature, winds, and wizards. Well, maybe not wizards. But human perceived time is not the same as relativistic time. It is not the same time that physicists had in mind when they accepted the definition of velocity as distance divided by time. To know they are not the same, all we have to do is examine our sense of how we move faster, drive faster, and work faster with impressions that suggest an overestimated marking of time. We really think that we can make up significant amounts of time when we drive for ten minutes at ten miles an hour over a sixty-five mile per hour speed limit when we are late for work. We think we can shrink time with speed. We think that we will be saving minutes, but reality checks using the physicists' definition of time tell us that we are generally just knocking off seconds. By this example, the reality is that only eighty seconds were saved. There is a significant difference between the physicists' definition, a definition that cannot be anything different for real physics, and the human perception of time.

If we want an answer to the question of what time is in physics, we must start with observing its behavior, because understanding what something *is* is often defined by what it does. Travel at a colossal speed relative to earth and time will pass more slowly than if you had stayed on earth. Travel far from earth to someplace with a much weaker gravitational pull than earth, such as the International Space Station (250 miles above earth), and time will run faster.

Imagine identical twins, one who travels at an enormous speed while the other stays on earth. We have the so-called clock paradox or twin paradox (take your pick), which was confirmed by twin American astronauts in 2016. Scott Kelly spent 342 days in orbit at the International Space Station traveling around the earth at about 17,100 miles per hour relative to the earth's surface. He returned home in March 2016. According to Einstein's theory of special relativity, which predicts time dilation, time for Scott had moved more slowly than time for Scott's six-minute-older twin brother, Mark (also an astronaut, the husband of former U.S. representative Gabrielle Giffords), who remained on earth for the time his brother was in orbit. Scott's voyage made him younger by about 28 microseconds each day; after 342 days at the station, he was younger by 8.6 milliseconds than if he had stayed on earth with Mark.

But hold on, maybe our cells are altered by our extended journeys and the speeds of our trips. NASA research investigations show that by being in space for 342 days Scott Kelly's body cells had experienced genetic mutations. It seems that humans in space do undergo certain biological changes. Of course, they age like anyone else; however, those astronauts who spend long durations (almost a year) in space undergo extraordinary hazards of confinement, isolation, microgravity, and radiation that alter gene networks and other immune-related biological variables. In particular, there are molecular physiological changes. Each end of a chromosome has what is called a telomere, a repeated segment of DNA consisting of an ordered sequence of three essential bases of nucleic acids: thymine, adenine, and guanine. Those end telomeres protect the chromosome from depreciation and from fusion with nearby chromosomes. Telomeres tend to shorten with age.[3] When they do shorten, the protection against radiation, environmental toxins, and detrimental emotions of stress is diminished. Unexpectedly, Scott Kelly's telomeres lengthened, quite possibly because his body produced new cells with longer telomeres as a defense against the harsh space environment that is so alien to human physiological endurance. If lengths of telomeres are statistically suggestive of marginal measures of age, Scott was actually getting younger the more he stayed in space!

This time dilation seems to defy common sense. While in constant relative motion, *each* twin appears younger to the other. But Scott ends up younger because the symmetry is broken when he accelerates to change reference frames for the return journey. But in physics the phenomenon is not at all a paradox. Each twin sees the other's time dilate, so each appears younger to the other. It is hard to fathom that time is not absolute—not simply because the physicists have made a definition linking space and time through speed and/or gravitation but more because it seems to imply a profound contradiction to what we understand as the biological aging process. After all, our cells are not speeding up in our bodies when we travel fast. Relative to our bodies, our cells are going nowhere.

We can look at the commonsense issue of this paradox in two ways. The first is to think about time in the universe logically, without any impression that mathematics gives to our senses. We sit in a room, perhaps the room you are now sitting in, and think that we are stationary in that room. Further consideration tells us that we are not stationary but rather moving at a frightening speed of one thousand miles per hour as the surface of the earth spins around its polar axis. Add to this the speed at which earth is moving along its orbital path around the sun at more than eighteen miles per second! We don't think about that and we don't feel that speed; yet, in the time you read that last sentence, relative to the sun, you have moved more than one hundred miles. Relative to any table or chair in the room, aside from the normal muscular movements of your body and moving your eyes along the page, you didn't move at all. Recognizing that we rarely think of ourselves as part of the bigger world of the cosmos gives us some sense of the wild relativity of motion, and hence of time.

Nothing can ever move faster than *the speed of light in a vacuum,* except, of course, the Starship *Enterprise* and other fictional vehicles that move in very peculiar media. That fact leaves us with some strange, unintuitive understanding of time and space, some wild notions that are far beyond any knowledge developed from our experience. You look at a ball. Its shape seems to be spherical. Someone else looks at the same ball from a different moving frame and it appears oval, or flattened in the direction of its relative motion. It begs the question of what we mean by "the same." Everything depends on the observer's frame of reference. This is the stuff of science, the stuff of phenomenology, not the stuff of human experience. The relativity of time and space, with all its experientially strange issues of dilation and counterintuitive frame transformations, is as real as what a clock on a wall has to say.

Surprisingly, accelerations of clocks relative to rest have no effect on time. The velocity of the moving clock is key. As velocity changes so does time dilation, but that dilation is dependent only on velocity.[4] According to an earth-bound observer's measurements, an astronaut's length shrinks while she's traveling away from earth at constant speed; when she lands back on earth, though, the earth observer measures her length to be unchanged, yet finds that that same astronaut is younger.[5]

The way to understand Einstein's twin paradox is to notice that our well-meaning physicists have, by their definition of time, put our sense of time's meaning into a corner. We think that our wonderful human time is physics time, when it is not. Mathematics has always given us universal truths, but only as abstractions from reality. Euclid's geometry works—at least it does in a locally flat space. It is, has always been, and always will be true because it is based on commonsense assumptions that must be universally agreed on by all who want to use Euclid's kind of geometry for its abstract implications or to build models of reality. It can be applied to the real world, but only applied. It does not tell us about all the nuances of reality that sometimes sneak up on mathematical models that do not account for unexpected generalizations, such as the non-Euclidean notions of curved spaces. If one looks deep, deep, *deep* into those nuances, there will always be something unexpected in the depths. It doesn't mean that anything is wrong or untrue. Rather, there are new angles to think about, new variables that were not needed at the shallower levels. That is why there seems to be confusion between the definition of velocity, which at reasonably slow speeds requires a link between distance and time, and the human notion of time, which is not so rigorous as to be blocked and cornered by physicists' equations involving the kind of human time that a simple clock can measure.

---

A clock must be able to count. To be a clock, it must interpret intervals of time as discrete ticks that can be counted. One possible model representation is this: Imagine a light reflecting between two parallel mirrors,

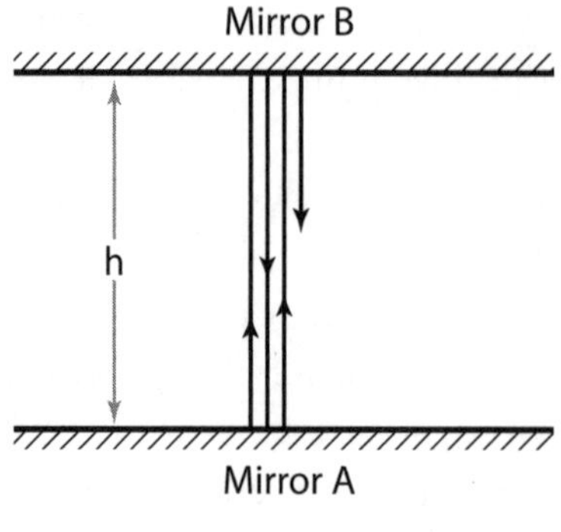


a

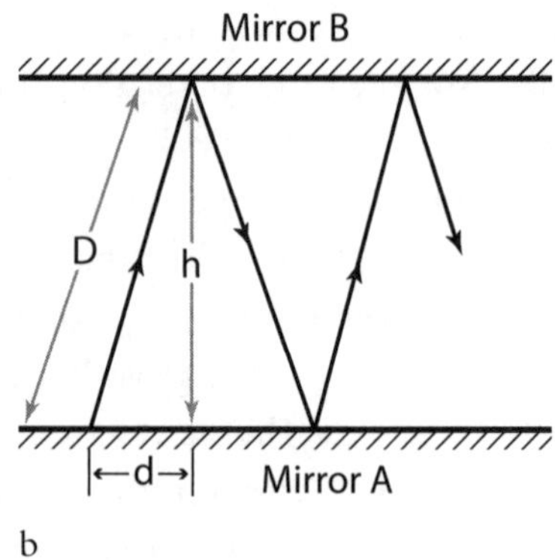


b

Rest clock (a) and moving clock (b)

A and B, as represented in the illustrations. To be a clock, something would simply count the ticks as back and forth reflections. An observer at rest with the mirrors would simply register the time it takes for light to go from one mirror to the other. However, if the mirror clock is moving to the right at some speed, then a stationary observer will see the path of light beam make a sawtooth pattern (7b in the illustration). We can see from the geometry that the light must travel farther in the moving clock. Because the speed of light is a constant in a vacuum, light takes longer to complete its back and forth reflections.[6]

So, *ah-ha,* this might suggest, at least from the physics perspective, that time *is* merely an illusion that depends on the observer and that there is really no such thing as some unique thing that we can label as *t* and call time. Perhaps we should think of time in its plural, *times.* We may have some convention that gives us an impression of what local time is. We look at the village clock and say, *Yes, that clock is telling me that I have to be where I am expected to be, but that's all.* That clock is simply a tool to organize my day.

Someone might say, *Wait! The mirrors could have a long enough length that the beams are just going up and down and not aiming for any particular spots on the actual mirror.* With an infinite mirror, neither the stationary observer nor the moving observer would notice the mirror moving. However, if a stationary observer with astonishing Godlike precision eyesight could see a light beam at all, he or she would see it moving at a slant between the mirrors in a sawtooth pattern. The best way to describe how someone with ordinary vision would see it is to think of someone moving while dribbling a basketball. The person dribbling sees the ball bouncing up and down perpendicular to the floor. A person standing still will see the ball's trajectory as a wave. He or she would not see a sawtooth wave because the speed of the ball changes with distance from the floor. The ball accelerates from the hand, reaches the floor, and highly decelerates when the skin of the ball reaches the surface of the floor. Both ball and floor are given energy from the ball as it deforms slightly while its velocity gets down to a zero. Both ball and floor quickly spring back to their normal form to repay energies back to the ball, which accelerates back to the hand of the player. Mechanical clocks act in similar ways with their escapements giving and taking energy with each hit of the hand and each bounce from the floor.

---

Einstein's "Principle of Relativity" was based on mathematical models, not hardened laws that could be proven by pure mathematics. Einstein raised relativity as a conjecture about motion in a system of standard Cartesian coordinates in Euclidean geometry while assuming that Newtonian laws of motion apply. To describe the motion of an object centered at some coordinate point in such a system the usual way is to give coordinate values as a function of the time variable $t$. So the coordinate of something traveling through three-dimensional $x$-$y$-$z$ space would be expressed as a group of three variable distances depending on time. The coordinate of that something at time $t$ would be denoted as $x(t),y(t),z(t)$. Einstein was very clear at this point to let us know something that most undergraduates don't think about when getting to this idea of using time as a parameter: "a mathematical description of this kind has no physical meaning unless we are very clear as to what we understand by 'time.'"[7] Without that understanding, all we are doing, when looking at motion through the lens of a parametric equation based on time, is dealing with an abstraction that has nothing to do with physics or metaphysics. Without considering what's at root here, we miss the critical point of the mathematical description of how real motion in the real world works. We have to consider that when we think of time—not just the variable $t$ that has no physical meaning, but time itself—we must think of it in the context of comparative simultaneous events.

We have no trouble accepting motion as relative. Watch two cars, one black, the other white, traveling side by side at, say, fifty miles an hour. Ignoring the road and background, you see both traveling at the same speed, fifty miles an hour. The drivers see each other as, relatively, not moving at all. If the black car were traveling at sixty miles an hour, the driver of the white car would see the black one as if it were relatively moving at ten miles an hour. The sensation is amplified in trains, where the sound of an engine is far from the passenger cars. A slow-rolling train car is quiet and smoothly rolling. We feel a disorienting sensation of confused relativity when we are in a stationary train looking out a window watching another train on a parallel track. When the train we are watching slowly begins to move, we feel an uncomfortable perception that our train is moving even though we know it is not.

We cannot tell which frame of reference we are in if it's closed off from outside. Therefore, one coordinate frame is as good as the other for deducing the laws of physics.

Yet we do have a problem. The simultaneity of measuring times. Is the now for one person the same for another person? Is the now of one clock the same as another some distance apart?

# INTERLUDE: TIME ON THE INTERNATIONAL SPACE STATION

Michael López-Alegría had three missions to the International Space Station (ISS) and was commander of Expedition 14, launched on the *Soyuz* from Kazakhstan in 2006. I asked him about his sense of time when traveling to the ISS and during his seven months at the station. I fancied that he must have been sleepless on arrival and that adrenaline was pumping from excitement.

"When I flew, it took about forty-eight hours to get to the station on a space shuttle and just six on the *Soyuz*. I didn't think about time on the shuttle, but yeah, I did on *Soyuz*. The shuttle kept us too busy to think about time. We were too preoccupied by all the details of our work with the payload and navigation along the way. The *Soyuz*, on the other hand, is a simple spacecraft with nothing in it and not much to do, so time on the *Soyuz* moves slowly. Of course, the excitement index is high, so your mind is preoccupied by the thrill of going into space. The adrenaline is running fast, and so even though you do get a bit bored on the *Soyuz*, you don't really think much about time."

López-Alegría was at the ISS from September 18, 2006, to April 21, 2007. That comes to 215 days, a long time to be away from earth. If one counts ninety-minute intervals of orbiting earth, 215 earth days translates to 6,880 swaps of days and nights aboard ISS.[1] Before speaking with López-Alegría, I had the impression that being on ISS must cause serious interferences with circadian rhythms, confusion for melatonin production cycles, and body-clock disequilibrium even after the routines of ISS are established. López-Alegría corrected my thoughts. On the station astronauts are always on Greenwich Mean Time. The lights were always on during the "day," off at "night." If you looked out a porthole window you could tell that the station is passing over earth's twilight or daybreak every forty-five minutes, but that doesn't have any effect on the people inside. They just keep to GMT for their senses of day and night, which was not at all dictated by the position of ISS. For López-Alegría, there was no interference with circadian rhythms, no confusion in melatonin production cycles, no meddling with body-clock equilibriums.

The team had designated eating and sleeping times. With all the business of the day it was easy to fall asleep at sleep time. The problem for López-Alegría was in staying asleep. He would fall asleep and wake two hours later. It might have been a melatonin issue; it might have been being strapped in a sleeping bag. Normal sleep needs a bit of natural movement. When you're strapped in you can't move easily. But you need to, and when you can't, you wake up. López-Alegría would wake, stay awake for a time, and then fall asleep for another two hours, often taking medication for sleep.

# 9

## SIMULTANEOUS CLOCKS (CALIBRATED TIME)

In 1898, seven years before Einstein's seminal special relativity paper, Henri Poincaré wrote a critical paper, "La mesure du temps" (The measure of time), asking what simultaneity is and suggesting that whatever it is, it must be defined and be relative to a person's point of view.[1] It was a lashing of old philosophical theories about intuitive perceptions of time, yet also an idea about relativity very close to what Einstein would discover just a few years later, that the clue to simultaneity was in the transmission of time between one place and another.

Although it is not clear whether or not Einstein had read "La mesure du temps," it is evident that he had read some of Poincaré's papers. "In Bern," he wrote to his close friend Michele Besso, "I had regular philosophical reading and discussion evenings. . . . The reading of Hume, along with Poincaré and Mach, had some influence on my development."[2]

In *Einstein's Clocks, Poincaré's Maps: Empires of Time,* the American physicist and physics historian Peter Galison asks: What was it about the turn of the twentieth century that brought Einstein and Poincaré to think of simultaneity in terms of coordinating clocks by electromagnetic signals? In a partial answer, Galison gives us this ponderous paragraph:

> Once in a great while a scientific-technological shift occurs that cannot be understood in the cleanly separated domains of technology, science or philosophy. The coordination of time in the half-century following 1860 simply does not sublime in a slow,

> even-paced process from the technological field upward into the more rarified realms of science and philosophy. Nor did ideas of time synchronization originate in a pure realm of thought and then condense into the objects and actions of machines and factories.[3]

In his 1905 paper, Einstein uses time as a variable without defining what it is. Where is that definition, aside from just naming time by the letter *t*? "If we wish to describe the motion of a material point," Einstein wrote in his essay "On the Electrodynamics of Moving Bodies," the values of its coordinates must be expressed as functions of time, and "We have to take into consideration the fact that those of our conceptions, in which time plays a part, are always conceptions of synchronism."[4] This, of course, is not a definition. The definition was buried beneath and appeared as a quasi-definition only by the implicitness of the time coordinate's locked linkage to the three space coordinates. The cam in the lock is the speed of light *c*, which both connects a spatial unit with a time unit and is constant in a vacuum. In consequence, time cannot do anything it likes to do without affecting space and vice versa. So Einstein tried to strengthen his quasi-definition by giving an example. He supposed that he is at a train station when the hands on his watch indicate seven o'clock and a train arrives at seven o'clock. The arrival of the train and the positions of the hands of his clock are then synchronous. He is therefore substituting the reading of the clock for time, but only for the space coordinates of the clock, not for places that are at remote distances from the clock.

Einstein might have thought that his example skirts the difficulties of defining time. The problem—yes, Einstein knew there was a problem—is that coordination of times at remote distances requires observers. It would be tricky to have observers watching and timing signals from one station to another through empty space. Fortunately, the definition could be eased considerably by assuming that the two stations being watched are both stationary—that is, both observers and things being observed are not moving. One observer would be at *A* recording time $t_A$ as measured by the hands of a clock in the immediate area of *A*. At a remote point *B* there would be another clock resembling the one at *A*. An observer at *B* would record the time $t_B$ of events in the immediate area of *B*. The two times $t_A$ and $t_B$ would be compared.

Is there a way to know a common time for both $A$ and $B$? To define such a common time there must be a way to measure the time it takes light to travel from $A$ to $B$, and that time must be equal in measure to the time it takes light to travel from $B$ to $A$. Therefore, consider a light that starts from $A$ at time $t_A$, arrives at $B$, is reflected back to $A$ at time $t_B$, and arrives back at $A$ at time $t'_A$. From this, we have a definition of synchronization, or at least a notion of when two clocks are in synch: Two clocks are *synchronized* if $t_B - t_A = t'_A - t_B$. In a strange sense, this circuitous definition of clock synchronization gives us the best definition of time in physics that we have. Notice that it is circular because it assumes that time itself is already understood by the measurements taken at $A$ and $B$. "Thus," Einstein wrote, "with the help of certain physical experiences, we have established what we understand when we speak of clocks at rest at different stations, and synchronous with one another; and thereby we have arrived at a definition of synchronism and time."[5] But what does Einstein mean by "physical experiences"? Arguably, he does not mean the instinctual experience of time that Immanuel Kant talked about as a priori knowledge awakened by our cognitive sense endowments, our intuition that is extracted and conceptualized from all our experiences. Rather, I believe, he uses the expression "physical experiences" to mean what the clock hands do and how fast light travels.

In this way we have given ourselves a definition of time, at least a time in physics. The *time* of an event in a stationary system is the marking of a stationary clock located near the event as the standard, with the condition that the clock is synchronous with a specified stationary clock. Going further, Einstein assumed that the speed of light in a vacuum is a universal constant defined by $c = 2d/(t'_A - t_A)$, where $d$ is the distance between $A$ and $B$.

Suppose that we observe a rod whose length is $l$, measured at rest. Place two stationary clocks synchronous in the stationary system at the ends $A$ and $B$ of the rod, one at $A$ and the other at $B$. Place an observer at $A$ and another observer at $B$, so that at each clock and at each end there is an observer. Now consider a similar setup, a rod traveling at speed $v$, now moving with respect to a stationary coordinate frame with a constant velocity toward increasing values of $x$. As before, let $t'_A$ be the time it takes a light ray to start at time $t_A$ from $A$ and reflect back after reaching $B$ at time

$t_{B'}$. Then, assuming the speed of light in a vacuum $c$ is constant for all observers, those for whom the rod is moving will find that $t_B - t_A = l/(c - v)$, and $t'_A - t_B = l/(c + v)$.[6] So the two moving clocks are not synchronous in this frame, since $t_B - t_A \neq t'_A - t_{B'}$. But observers moving along with the rod and clocks (so that for them, $v = 0$) would see the clocks as synchronous. This means that there is no absolute concept of simultaneity and therefore no absolute concept of time, a concept that is, as we have noted before, strongly and strangely counterintuitive.

Einstein's time is mathematical time, not the time that humans grow up with, a synchronization of life events with the turning and orbit of the earth. Humans experience life by where the sun is in the sky, by their expectations, by their cares, and by their memories. In some sense, synchronization is time. The clock we know is a measure of the day we experience and the accumulated memories of events in an ordered sequence, even though we often get confused about the sequential order of those memorable events. Einstein believed that he had settled the definition of time, effectively through a clock at rest in a stationary system.[7] In other words, time itself is not really defined, but the time of an event is. The time of an event is the simultaneity of the event with the event of a stationary clock that is synchronous with a specified stationary clock. So, we are left to assume that the clock's event is the indicated time as represented by the positions of the clock's hands. The numbers those hands point to are irrelevant, except that they provide a numerical chart of their sequential positions. That kind of time, interpreted by the clock and by all synchronous clocks, is, however, some number denoted by the letter $t$.

How different this is from Newton's view of time. For Newton, time and space were absolutes, universals that were obeyed independent of an observer's motion. In Newton's physics, an observer should see light travel from an approaching source more quickly than if it were beamed from a stationary source. Time would then have two sides. One would be an absolute universal, a mathematical time independent of any outside reference, a time flowing uniformly in equal increments of duration. The other would be a common apparent time of the senses, a relative time that could have a precise or imprecise measurement by observing motion, such as the earth in orbit, giving us the hour, day, month, and year.

My five students were well prepared for this problem with an elementary understanding of physics. But they naturally continued to wonder about how all this mathematical abstraction maneuvering just happened to fit with the physical laws of the universe, particularly regarding Maxwell's equations. Yes, they knew about modeling and about how physicists and applied mathematicians often build ideas and equations to fit marvels of the real world, to express so much worldly phenomena as simply mathematics. They wondered how this abstraction could have connected with the physics of electromagnetic processes dictated by Maxwell's equations and other concepts of electron theory. This one abstraction seemed too incredible to them, and that was where their comprehension collapsed. Their thoughts were entrenched in their preestablished confirmation biases governed by environmental impressions of a Euclidean nature, as if the world is just some giant machine of connected gears, pulleys, and levers propelled by some metaphysical driving force. But this is the nature of abstraction. Physics builds its models of real life as components of a mathematical erector set that can be figured and reconfigured to indicate abstract truths that reflect back to hugely insightful understandings of the real world in which we live. Sometimes it just takes mathematical scaffoldings built from high school ground levels.

Like Newton before him, Einstein never really defined time in any other way than how a clock measures that indefinable thing we are calling time. Does time slow down under the influence of speed and gravity, or is it the clock that slows? It's a reasonable question. It seems that the clock itself defines time. By that I mean: it is the clock, a counter that ticks off units in a totally uniform, consistent, and continuous way that depends on where it is and how fast it is moving. We now, however, do have the physicists' notion of time, a clock that depends on speed, place, and observer. Does it tell us anything about past, present, and future? *No!* According to Einstein, "People like us, who believe in physics, know that the distinction between past, present and future is only a stubbornly persistent illusion."[8]

An illusion? What could Einstein mean here? He is not talking about time here, just the temporal human illusion that there is a past, present, and future. He is suggesting that the past, present, and future are happening all together; they cannot be happening at different times. He might have meant, by virtue of the relativity of simultaneity, that one event may be in one person's future and also in another's past. Can anyone reading this book "now" compare what is happening in that same "now" to someone on the other side of the world? No, because that "now" had come and gone the instant it supposedly passed from the present to the past. But maybe the present and past are illusions, sparked by memories, and maybe the future is prompted by our fears and hopes. And maybe the present—that indescribably small instant, that point we imagine to be on a mathematical timeline—is simply an infinitesimal snapshot of our own awareness, our consciousness telling us that we are alive. The future for us might be determined by our instincts for survival, but the present is the now that lights our consciousness so that we can see where we've been and where we're going. It is that sly, complex part of what goes on in the mind when thought, stimulation, and sensation are bound together whenever we are awake, and free to think.

But here is why Einstein wrote to his good friend Michele Besso that the distinction between past and future is a human illusion. What is the "now" of UDF 2457, a red dwarf star on the far side of our galaxy? If we imagine its "now" in earth time, it happened about fifty-nine thousand years ago. It's a mind-boggling thought, because we are confused about the simultaneity of its time with ours. But we needn't consider a place so far away. Just call Tokyo from New York. You might assume that you are speaking to a person and that his or her "now" is the same as yours. It isn't. We are not talking about the difference between the time in New York and the time in Tokyo. We've known for more than a century that time depends on both speed and gravitation. So, there will be a very small difference, no greater than a minuscule fraction of a second, in time between you and the party you are speaking with in Tokyo. However, Tokyo and New York are somewhat bound together in time because those two places both happen to be on earth. UDF 2457 is not linked to the earth and is moving quite fast relative to our planet. So its "now" instant will be very different from ours. "Now" is a personal instant, an instant that is different from the person sitting next to you. This does seem like a crazy, nonpractical philosophical messing of the mind, but it is real.

What do we learn about time from relativity? We know that the speed of light must be the same for all observers and that anything that has mass cannot travel faster than the speed of light in a vacuum.[9] The theory of relativity does not exclude other "things" from going faster. For instance, energy transfer and causation are not restricted by any speed limits. Take an electron and a positron that happen to be in what quantum physicists call a *singlet state:* two particles that originated from a single quantum event involving a single charge. They are linked by the condition that their net angular momentum is zero. These two "things" can travel in different directions and could soon be millions of miles apart. And yet, when we measure the spin of the electron, wherever it is, that measurement will instantly change the spin on a positron (a counterpart electron with opposite charge) on the other side of the planet. You might wonder how the electron can send instant information so far and so fast to a positron. It does so at a speed infinitely faster than the speed of light! Or so it appears. There seems to be some sort of synchronization of operations between two objects that are not local to each other. Einstein knew about this theory that is now dubbed quantum entanglement; with skepticism he said, "Physics should represent reality in time and space, free from spooky actions at a distance."[10] The surprise is that there is some exchange of information, at least quantum information, but not energy transfer. This is the spooky thing that we have trouble getting our heads around. When we question the nature of that information, we do not get any understandable answer, at least not information graspable to someone who comprehends physics and causal relations by classical reasoning. At some point of one's reasoning, one must ask if there can be any interconnection phenomenon between two distant elements that permit a marvel of signals traveling faster than the speed of light. If there can be, we must rethink the meaning of synchronicity, and therefore of time. But we are not quantum particles!

So, it questions the question. We might not really know what time is, yet we do know ways of measuring its passing, whatever it is. When physicists use *t* in their formulas and find all sorts of odd things about its relativity, all they are saying is: take your pick of units for measuring time, and then let's call that measurement *t*. Whatever it is, we have the mathematics relating it to motion that tells us that all those strange things happen. We don't care how you measure the *t* that we use in our formulas. Let it be the beats of the king's pulse or the setting of a metronome. Whatever *t* is, the physics will dictate the same result.

## INTERLUDE: ANOTHER TIME ON THE INTERNATIONAL SPACE STATION

When I interviewed Samantha Cristoforetti, an Italian astronaut from the European Space Agency, I got a story different than López-Alegría's. She, too, was transported to the station on the *Soyuz* from Baikonur in southern Kazakhstan in 2014, and she stayed on the station for six and a half months.

"I have a rigid body clock," she told me. "The launch was at 2:00 a.m., and docking time at the station would have been a bit more than six hours later. Even though the adrenaline was pumping, there were intervals of fighting being awake and then intervals of trying to stay awake. All along, I was adjusting to weightlessness. While *Soyuz* circled around the earth trying to catch up to ISS, I was very alert watching the computer."

"How did you adjust while on ISS?"

"As I said, I'm a good sleeper and my sleeping arrangement on ISS was comfortable. And yet the first few nights my body clock was off, but eventually I could sleep at the designated sleep time. After one month on ISS, looking back, I had the feeling that I had been on station for a longer time already, that my life on Earth was far away."

I asked her if the confinement to a small space had some effect on how time passed, but she claimed that there was no confinement. In a room on earth you can move around in only two dimensions as you walk along its floor. A room in the station is three-dimensional. You're not restricted to move in just two dimensions, so it feels like plenty of room. The mathematician in me had not thought of that third degree of movement and how that extra dimension opens a new notion of comfort in movement. Unlike the differing stories of long-haul truckers, all the astronauts I interviewed who had extended times in space had consistent stories.

# 10

## BRACED UNIFICATION (SPACE-TIME)

In 1907 the German mathematician Hermann Minkowski developed a geometric representation of four-dimensional space-time. "Henceforth," he said on September 21, 1908, addressing the Congress of Natural Philosophers, "space for itself, and time for itself shall completely reduce to a mere shadow, and only some sort of union of the two shall preserve independence."[1] He saw the union as a liberation of established fixed time and space as connected to his geometrical formulation of four-dimensional space-time. Space and time were no longer to be separated in physics; they were to be thought of as fused in an inseparable union. That fusion was critical to what would soon give a causal physical sense to how gravitation plays its role in the four-dimensional geometry of space-time.

"Subjects of our perception," he continued, "are always connected with place and time. No one has observed a place except at a particular time, or has observed a time except at a particular place. Yet, I respect the dogma that time and space have independent existences."[2] His world was the one-world manifold system of all possible values of $x$, $y$, $z$, $t$, an abstract description of everything that takes place in the universe. In every place and time, something perceptible happens to some matter or substance. With the symbol $dt$ considered as an element of time duration (some very short interval of time), there are correspondingly small elements of space $dx$, $dy$, $dz$, extending a spatial increment in three dimensions. In this way we get a picture of the life of the point $(x, y, z)$ on a curve in a four-dimensional world. "The whole world appears to be resolved in such world-[curves]," he wrote, "and I may just anticipate, that according to my opinion the physical laws would find their most perfect expression as mutual relations among these world-[curves]."[3] The link between time and space comes from the axiomatic understanding that $dx$, $dy$, $dz$, and $dt$ are not independent but rather are linked by the physicist Hendrik Lorentz's equations of length contraction, and physics' consequent agreement that the speed of any point of matter $v$ cannot be greater than the speed of light $c$ in a vacuum, and therefore by the strict inequality $c^2dt^2 - dx^2 - dy^2 - dz^2 > 0$ that must be obeyed.[4] Such a condition connects the time dimension and space dimensions, distinguished by opposite signs in the dimensions of time and space. So, although space coordinates may be thought of as being independent of one another, the time coordinate enjoys a relation involving all three space coordinates.

The magic of Minkowski's idea of space-time is hidden in the fact that the speed of light in a vacuum is constant. It might also be the clue to why we find it so difficult to understand what time is, if it is anything at all. Space-time is an oxymoron in the sense that it is unusual for a geometrical coordinate system to mix units. The first three units are in units of distance, while the fourth looks as though it is a unit of time. Of course, there is a way around that by converting the time unit to a distance unit. Since the speed of light in a vacuum is constant, we can convert any time unit to light-years or light-minutes or light-seconds. One light-year is the distance that light travels in one year. If you make a reservation in a restaurant for 7:00 p.m. and it is just 5:15 p.m., then you could just as well say that the reservation is for 0.00000402624 light-seconds from now. This is, of course, silly for a restaurant reservation, but it is very convenient as a measure of astronomical distances, when light travels at a speed of 186,300 miles in just one second.

At the fault line of this paradigm quake is the understanding that space-time intervals are absolute, independent of the observer, even though space and time separately are not. It means that two observers moving toward or away from each other draw different conclusions about time, about when things will happen and have happened.

With this, the physical universe becomes totally spatial in all four of its dimensions. "The world," according to classicist Eva Brann at St. John's College, "would become absolute, 'pre-written,' and the observer would

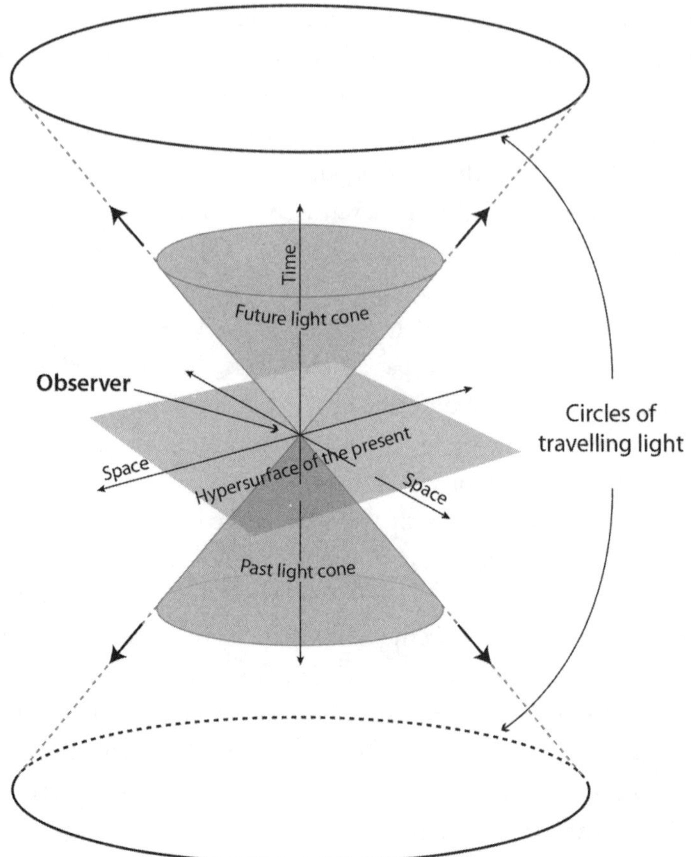


Light cone of now and then

waft through it from perceptual moment to moment, experiencing as Becoming what was really his particular intersection with Being."[5] And with this we find every human observer passing along his or her personal world line. It all becomes geometry, a mixed-up word whose seeds belong to earth, with a meaning that extends across the entire universe of human imagination. Imagining Minkowski's world, we find the here and now of the present time of any observer at the origin of its coordinate system. The geometry of space-time gives us a cone centered on that present "now," one side opening to the future, the other side toward the past.

Imagination can take us far. However, to visualize the four-dimensional Minkowski world, we must pare the four dimensions down to three. Imagine a two-dimensional plane with light originating at its origin and traveling in three dimensions, the third dimension being the time coordinate *t*. Because that light has no preference as to where it will go and because its speed is believed to be constant, it will travel in every direction in three dimensions at once. At any fixed time *t*, the light will fill a circle.[6] Therefore, the traveling light will fill a cone in three dimensions, the forward movement following positive increasing *t* and the backward movement following increasingly negative *t*.

Along with this description of space-time, coupled with general relativity (which extends special relativity to include gravity), comes the prediction of black holes in the fabric of the universe. These are not regions of space-time smoothly dimpled by finitely heavy masses; rather, they are smooth, sharply curved regions with such strong gravitational effects that nothing can escape—not things, not light, and not even time. The gravitational force nearby is so strong that it sucks in everything in its reach. Inside, there is a perpetual now—no past, no future. Get near one and you must move faster and faster to escape its stronger and stronger gravitational pull. And just now, at the time of the writing of this book, for the first time ever, we have images of a supermassive black hole at the center of the galaxy M87 (55 million light-years from earth).[7] With more than a dozen radio telescopes around the world looking for holes, who knows what will come?

# INTERLUDE: A CURIOUS DIALOGUE

The twin paradox, which is not a paradox at all, seemed to be mandatory for any sophomore physics class talk about relativity, since it was always an attention grabber and conversation starter. So there we were, talking about the twin paradox, personified by the story of the twins Scott and Mark Kelly in chapter 8, the special relativity thought experiment considering identical twins, one an astronaut who wanders around space before returning to earth to find his twin brother more aged than he. Soon, our conversation veered off from discussing the age of twins to a conversation of what an astronaut might actually think about during his journey out of the galaxy. My student asked me to imagine him as an astronaut on a mission to outer space, flying at half the speed of light, taking him fifty years to get to wherever he is.

"What does that mean?" he asked.

"What does what mean?"

"Fifty years. My watch says fifty years. If I look back at earth and count the number of times the earth orbited around the sun, that number must be 50. I mean, the earth doesn't slow down just because I'm moving at such a great speed. Either it's orbiting around the sun at some speed or it's not. What do I have to do with changing time on earth?"

I had to think for a moment. It was a beautiful observation that gave me a wonderful opportunity to clarify what it means when physicists talk of time as being *dependent* on the observer.

"Yes," I said. "The pace at which the earth travels on its way around the sun will appear slower to you than to earthlings on earth. The number of orbits the earth makes around the sun will be less than the years that go by on earth—to you!"

"That means I am affecting something happening elsewhere!"

"It would seem as if you are affecting something happening elsewhere, but you aren't. Earthlings will see things differently than you do. And that is the whole point of relativity."

"What is it that makes me see things differently?"

"You are traveling at a colossal speed, right?"

"Yeah, 93,141 miles per second!" (It seemed that he thought his question through.)

There were a few ways to answer my student. I took the view (writing with chalk on a blackboard) of an earthling's time interval as $\Delta t'$ and his as $\Delta t$, with a relation given by $\Delta t' = \Delta t/\sqrt{1 - (v^2/c^2)}$. Therefore, at his imagined speed $\Delta t' = 2\Delta t/\sqrt{3}$. In other words, from an earthling's point of view, $\Delta t' > \Delta t$. But that doesn't give the whole reason.

"Look at the earth orbiting the sun one day," I said. "Mark its position on its orbit."

"How do I know what a day is?" he countered. "I'm out in space. There are no days in space. And for that matter, I don't know what an hour is."

I told him that he seemed to forget that he has a very precise astronaut watch that was set to Greenwich Mean Time before he left. It would still be counting at a frequency of cesium-133. But at his speed of half the speed of light, that frequency would have appeared to be slower to a person still on earth. He could look at the earth orbiting the sun, then look again one year later—one year, going by his watch. He might think that the earth completed its orbit around the sun. But from the point of view of a stationary earthling, more than a year would have passed.

"Why is that?" he asked.

"Because the earth that you see is no longer in the position that you see, because the light has taken a great deal of time reaching you. At your speed, it's hard for light to catch up. So, what you are seeing actually happened long ago. Where you see the earth on its orbit is not where it really is."

"What do you mean by really?"

"Exactly!"

# 11

## ANOTHER MIDNIGHT IN PARIS (TRAVELING THROUGH TIME)

> Oh, the attic's a dark, friendly place, full of Time, and if you stand in the very center of it, straight and tall, squinting your eyes, and thinking and thinking, and smelling the Past, and putting out your hands to feel of Long Ago, why, it . . .
> —*Ray Bradbury, "A Scent of Sarsaparilla"*

In Woody Allen's film *Midnight in Paris* the protagonist, Gil, travels through time back and forth in a yellow Peugeot limousine time machine between present-day Paris and 1920s Paris, where he meets the legendary literati of that golden age. Back in twenties Paris, Gil meets Adrianna, one of Pablo Picasso's many presumed lovers (fictional in the movie), and, by horse-drawn carriage, travels back further to La Belle Époque of the late nineteenth century.

> GIL: Because if you stay here and this becomes your present, sooner or later you'll imagine another time was really the golden time. . . . The present has a hold on you because it's your present and while there's never any progress in the most important things, you get to appreciate—what little progress is made—the internet—Pepto-Bismol. The present is always going to seem unsatisfying because life itself is unsatisfying—that's why Gauguin goes back and forth between Paris and Tahiti, searching.[1]

In literature and in movies we go back and forth in time with ease, whether the machine is a Peugeot Type 176 or the Starship *Enterprise*. In

Stephen King's novel *11/22/63*, it is a porthole door in a diner. In the film *Back to the Future* it is a DeLorean motorcar, in the long-running TV series *Dr. Who* it is a telephone booth, and in H. G. Wells's *Time Machine* it is a . . . well, we don't quite know exactly what it is. Fictional traveling in time seems as easy as stepping into an elevator and pressing a button. The door closes at one place and very soon opens at another. Traveling through time, though, is not the same as traveling through space. The metric distance from one place to another is symmetric and fixed. Time does not have a fixed metric independent of motion, and it is certainly not symmetric with regard to motion.

For traveling back in time with just a one-way ticket, it's the attic for William Finch in Ray Bradbury's short story "A Scent of Sarsaparilla." "Well," Finch says to himself, " . . . wouldn't it be interesting if Time Travel could occur? And what more logical, proper place for it to happen than in an attic like ours, eh?"[2]

Can anyone travel back in time? After all, we can go back and forth on almost any journey in each of the three dimensions of space, so why can't we do that with that fourth dimension, time? Such a journey might not be practical, but is it possible? Mathematical physics tells us that it is, at least if we can travel on what physicists call a closed timelike curve (CTC), those world lines that are closed in space-time. Follow that curve and you return to where you are, but strangely you get there at an earlier time. Yes! I'm amused by the thought of some time-traveling visitors being among us. It is mathematically possible, but only mathematically possible. It would be fantasy to go back to a time before being born, to, say, 1938, before Kristallnacht, to kill Hitler. I would hate to go so far back as to be in high school again. But to kill Hitler and still be able to return to my own life with my wonderful marriage, my two wonderful children and five grandchildren? You bet! I would do it, if only I had a lion's courage.

As far as I know, none of the physicists who work on the theory of time travel is building time machines or rockets that can bring a human to a wormhole. Even the DeLorean DMC-12 that was the featured machine in *Back to the Future* is no longer being manufactured as a new car. Those physicists, the ones who work on time travel, do so for getting to the heart of the theory, certainly not for investments, and certainly not for any schemes of soothsaying. Such a theory must bring with it a sweeping current of ideas, and many of those ideas must also lead to new ones and to understanding the fundamentals of space-time, how the universe works, how it has always worked, how it began, and how it will end, if indeed it ever will.

If we view time as a fourth dimension—as physicists do—then, in the mathematical equations of physics, time is reversible, suggesting speculation that the grandfather paradox is real, the paradox that tells us that one could in theory go back in time with the intention to kill one's own grandfather and alter the possibilities of one's own birth. But those mathematical equations are mere models designed by physicists to give as much information about the real world as they can without disrupting life as we know it. It might be theoretically possible to reverse time mathematically, but the theory does not support any manipulations of life in the past that alter the future.

Many of the physicists I know view time travel as suspiciously dubious, since the entire concept permits effects to happen before their causes. Their wariness comes from knowing that general relativity equations of space-time curvature are local, that is: respecting all properties of space-time curvature in any small neighborhood surrounding any point in space, including the causality constraint backed by Stephen Hawking's *chronology protection conjecture:* cause must precede effect. Hawking tells us, "Euclidean wormholes do not introduce any nonlocal effects. So they are no good for space or time travel."[3]

Reversing time is physics, and mostly mathematics, but our experience tells us that time moves (if we can say that time "moves") in one direction only. The prophetic "time's arrow" points forward, not backward. So, there seems to be a dichotomy between time as defined by physics and time as defined by experience. Any solutions of a math equation involving time that obey the laws of physics are just as valid for time flowing backward as they are for flowing forward. The time variable can be moved forward or backward as easily as sliding two fingers on a touchpad to the left or to the right. We always get back to where we were without any possibility of changing one's fate. We may lose a tiny bit of aging in the process, but whatever happened along that time-traveling journey would remain the same. A dream can put us back in time, but we do awake forward in time, long past the moment we fell asleep. Awake, we are stuck in the present, no matter what we do, unless we agree that our memories can give illusions of traveling backward or forward along that single dimension of time on which space is dependent.

Physics also admits a more innocuously reversible time, one that swaps the future with the past. It turns logic inside out to leave us with reasoning that can happen only in the world of Lewis Carroll's White Queen in *Through the Looking-Glass,* who remembers the future but not the past. She is not time traveling. Her time forever runs continuously backward. "What sorts of thing do *you* remember best?" Alice asks the White Queen. "Oh," the Queen replies, "things that happened the week after next." She speaks while placing a large bandage on one of her fingers. Soon after, she screams and shakes her hand. "Have you pricked your finger?" Alice asks. "I haven't pricked it *yet,*" the Queen says, "but I soon shall—oh, oh, oh!"[4] The effectiveness of this dialogue shows that the White Queen's world flips our logical order of time to make us think about the consequences of our own. The humor makes us think about the reasoning, and the reasoning makes us think more broadly about our experience with time's arrow. Wonderland words have wondrous meanings.

For most of us who are already living in three dimensions, a fourth is hard to picture. However, that third space dimension, *up,* doesn't do much for us as we navigate around the country using GPS coordinates. So why not simply eliminate the "up" coordinate? If we do that, we might get a good sense of world lines by picturing a three-dimensional simplification of space-time. For every duration (take one year as a stand-in for my word *duration,* if you wish), the earth's world line is a curve in the three dimensions $x$, $y$, and $t$, where $x$ and $y$ are the coordinates of earth's movement in the celestial ecliptic and $t$ is the varying time between the start and end of that moment. A rough picture of that curve is illustrated here—rough, because we must remember that metric constraint $c^2dt^2 - dx^2 - dy^2 > 0$ linking time to space. You could picture an elliptical cylinder with a radius of (on average) 8.3 light-minutes (92.96 million miles, the distance from the earth to the sun). Earth's world line then spirals along the length of the imaginary elliptical cylinder. The sun too has its world line that could be represented as a straight line, if we consider everything relative to the sun. But the sun has its own world line. It too moves relative to other stars in the universe. So, our image of the cylinder on which

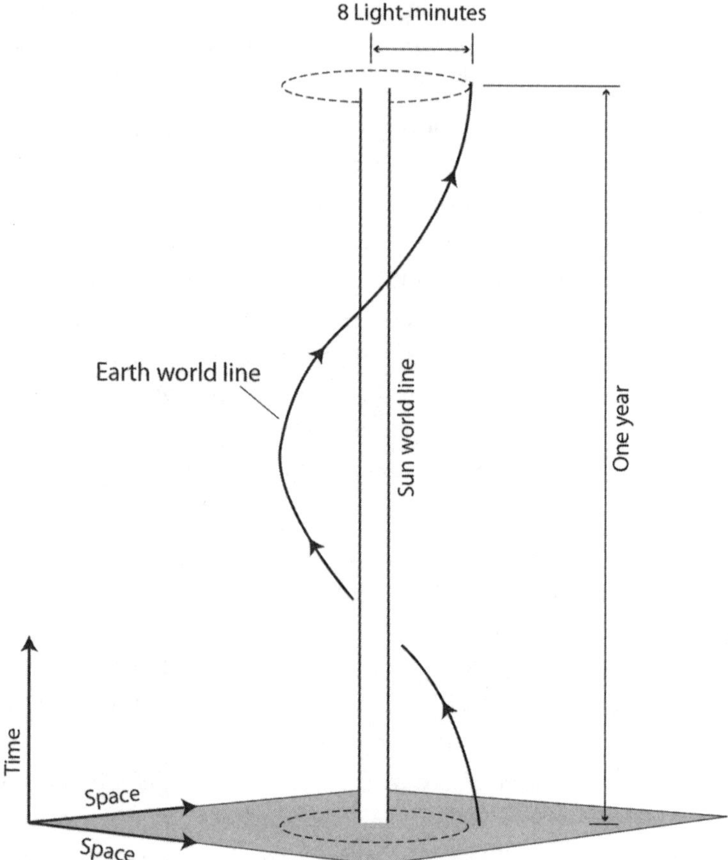


Earth world line

the earth spirals should be either a bit wiggly or bent to one side if not straight, even though *straight* has its own meaning in the universe.

So, what we actually see at any one instant of time—if we were watching the earth sail in its orbit around the sun—is a slice of the elliptical cylinder by a flat plane cutting through the cylinder. We see the earth and the sun as points on that cutting plane and can imagine the earth's orbit around the sun as an ellipse on that flat plane. We cannot see our own world lines, nor can we see our own histories as we move along those lines. Though they are more metaphorical than real, we do move through them from the moment we are conceived until the moment we die—even long after. We don't see ourselves as moving through a coordinate system in three dimensions of space. We see that coordinate system only because we were trained to see it when we learned about our geometry from mathematicians and physicists who not long ago devised these orderly envisaging factors and features. There was once a time when we did not envision ourselves in a world with three independent coordinates. This is a relatively new world, where math and physics show us things that we never dreamed of picturing. Indeed, our GPS world brings us back to a time when we had simpler thoughts about moving in just two dimensions. Ask your Google Maps app for directions from point *A* to point *B*, without any care whether *B* is higher or lower than *A*.[5] Your app draws a curve from *A* to *B*, shows you the distance between them, and tells you the time it will take to get from *A* to *B*. It will measure mileage by the road taken. This means that all steep inclines will be built into the app's algorithms, which in turn are based on what the road surveyors reported when the road geometry was last surveyed.

If we agree that we now live in a GPS coordinate frame, we can then see that a person's world line can be visualized as a curve in three dimensions coordinatized by three independent orthogonal (perpendicular) variables, $x$, $y$, and $t$. That visualization is just a shadow of the real four-dimensional curve, but that is all that is needed to follow a world-line experience. As the shadow moves through time, it shifts through space. Of course, life is not that simple. A person's experience is not like our naive image of the earth spiraling around the sun's timeline. A short period of a single human life is a braided entanglement of world-line curves: a person's food choices, love encounters, holiday travels, business trips, visits to grandma, books read, breaths taken, and the millions of experiences both small and large that happened in that short period. One could think of it as a group of world line strands braided to form one whole world-line curve, what the twentieth-century Russian American cosmologist and popular science writer George Gamow called a "world-band." This is how Gamow saw space-time geometry: "The topography and the history of the universe fuse into one harmonious picture, and all we have to consider is a tangled bunch of world lines representing the motion of individual atoms, animals, or stars."[6]

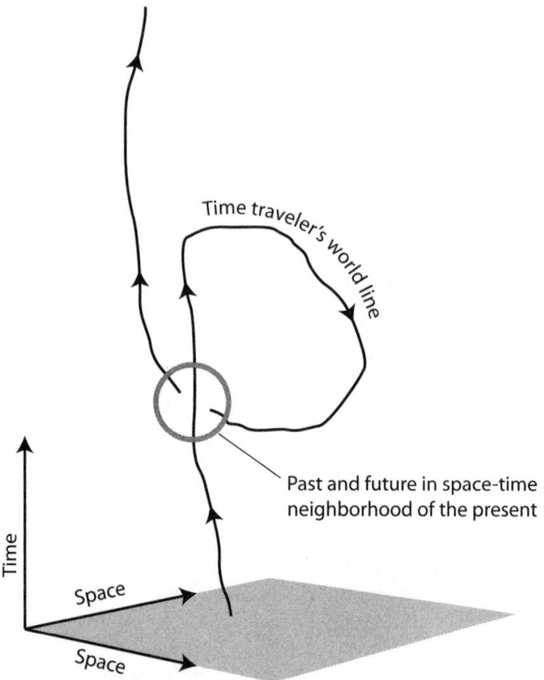


Looped world line

The illustration here represents the lucky time traveler's convenient loop of a world line, a CTC that loops in such a way that a *now* meets with a *past*. It seems to be a traveling jump in time, yet it is just a meeting. *Now* is still now and *past* remains in the past. It's a fine picture, but could a timelike curve ever meet itself? Stephen Hawking, by his *chronology protection conjecture,* speculates that a finite CTC would refute some laws of physics. "The laws of physics do not allow the appearance of closed timelike curves," Hawking wrote in 1992. Evidently, such a curve would have to be infinite, thereby forbidding any kind of normal, finitely scaled time travel through looped timelike curves. He quipped, "It seems that there is a Chronology Protection Agency which prevents the appearance of closed timelike curves and so makes the universe safe for historians."[7]

Those timelike curves that touch themselves in finite loops support time travel opportunities in the genre of science fiction, where jumps into the past or jumps into the future are almost always soaring pole vaults. It's confusing. The nonfictional astronaut who travels at fantastic speeds for a year comes back to earth younger than if she had stayed on earth. She is not younger than she was when she started her journey. She has not time traveled. She has traveled into the future, but in reality all that had happened was that two clocks—hers and earth's—were scaled differently. If she wandered the galaxy at a practically impossible speed very close to the speed of light for one second before returning in the next second, she would have returned to earth perhaps a decade later in earth time. But she had not traveled *through* time *into* the future. Her time had contracted relative to earth's time. No time was ever skipped over.

In the real world every experience is connected to places and times. We tend to separate our perceptions of space and time, partly because we learned them in separate experiences. However, we should not be thinking of space and time as separate entities, just as we should not think of letters on a page as separate from words or musical notes separate from timing. We all travel along *with* time at different rates. One person's appearance of aging moves faster or slower than another's.

The confusion comes when we consider the difference between time $t_a$ for the astronaut and time $t_e$ for someone on earth. Time $t_e$ is the common time calibrated to the motions of our planet and synchronized to all local stationary places on earth. As the scale moves forward for earth time $t_e$, so does the scale for the astronaut, just more slowly, always maintaining the inequality $t_a < t_e$.

The astronaut who returns to earth to find all her friends so much older than before didn't jump ahead in time, though she might feel as if she had. Her time foreshortened to make things feel as if there was a jump. Blame it on the wonderful effects of good science fiction that reinforce the confusion with swaps of temporal moments when a person instantly leaps through a magical porthole from the present into a date far in the future. The science fiction assumption is that there is a past, present, and future out there somewhere and that the past of our world is actually the present of another. Gil's trips to the Golden Age and La Belle Époque have multiple interpretations—that's how good fiction works. Perhaps there are simultaneous worlds, alternative universes that are moving through time in a manner that is out of synchronization with our own. If our astronaut could get to them somehow, she would have time traveled; otherwise, her trips to outer space are merely travels through time at time scales different from those on earth.

The twin *Voyager* space probes launched in 1977, back when recording technology at NASA was done on reel-to-reel tape recorders, entered interstellar space in 2012, and are still transmitting signals back to earth as of this writing. Now traveling at a speed of 9.6 miles per second, they will continue on a journey to nowhere in particular for hundreds of thousands of years (in earth time) or until they are found by some improbable intelligent beings who live by clocks very different from our own.

Even on earth we all live by different clocks. In a way, we are all living in different time worlds. The twin paradox in chapter 8 gave an account of twins, one of whom traveled around earth at about 17,100 miles per hour. Any speed will do. Think of time worlds this way. Two people, named for the sake of referral Amy and Bill, are given amazingly precise identical watches. They synchronized their watches to ding after precisely one hour. Amy rides a horse on the Central Park carousel while Bill stands in one place to the side of the carousel, looking at the rider. Bill will hear his watch ding before Amy hears hers. Time stretched for Amy, albeit such a tiny stretch that it would take a clock with accuracy better than thousandths of a billionth of a second to tell that there is a difference. During that hour, Amy and Bill were living in two different time worlds. In fact, by Einstein's theory of special relativity, almost all living things are living in different time worlds.

This suggests that Bill's lifetime depends on his movements during that lifetime. He sleeps in a bed, works with his laptop on a couch, spends summers reading newspapers on a beach, has long meals in restaurants, and doesn't get out much during the day. Amy walks her dog three times a day, goes for a jog every morning, commutes to work for an hour in each direction, and, as part of her job, once a month flies from New York to San Francisco. In comparison, Amy's aging is slower than Bill's. Granted, the rate of change of her age, even over a lifetime of consistent lifestyles, is hardly measurable.

A carousel might be a slow test of relative time, but in 1971 two physicists, Joseph Hafele and Richard Keating, tested this idea by taking synchronized cesium atomic clocks on commercial airliners for two trips around the world, one eastward, the other westward, to compare their times with stationary clocks in the U.S. Naval Observatory in Washington, D.C. In the end, the clocks disagreed with each other by a difference of as much as 59 nanoseconds, proving that a moving clock runs more slowly than one at rest. Complementary to the Hafele-Keating experiment is the 1959 Pound-Rebka experiment that tested the effect of gravitation on time as predicted by Einstein's theory of general relativity and found that in a gravitational field, clocks at different places run at varying rates. Clocks run more slowly under increased gravitation, so clocks at higher altitudes run faster than clocks on the ground.

These experiments certainly destroy Newton's theory of absolute time. Everyone moves at a different speed; a clock on a mountain ticks faster than one on an ocean; and a clock in space, away from huge masses, will tick faster than a clock on earth. Time on the 102nd floor of the Empire State Building is shorter than time in the lobby, and shorter than time even on the 101st floor. Of course, we are still talking about much smaller than nanosecond differences, likely smaller than even picoseconds (1/1000 of a nanosecond). However, by extreme comparison, approach SN 1054, a neutron star chronicled by the Chinese in 1054, and your lifetime is expanded considerably. SN 1054 is so massive that its gravitational attraction slows time down by almost 30 percent of time on earth.

Suppose I want to live a bit longer, say a hundred years longer, so that I can have the unique human luxury of knowing what my offspring will achieve, and perhaps attend one of my great-great-great-great-grandchildren's college graduation. All I'd have to do is take a wander around space, wherever I wish, at a high speed of fairly close to the speed of light (actually 99.9992 percent of the speed of light, or 186,281 miles per second). Once I get beyond the nauseatingly painful gradual acceleration to get up to goal speed, and then cruise at that speed at a comfortable constant velocity, I will hardly sense moving. After the joy of peeking at a few dwarf stars and returning to earth, time will have moved on by a hundred years. I'd likely be amazed to see self-driving cars levitating on roads that don't look at all like roads. I might see a few very old people jogging to work with prosthetic legs, see brainwave bonnets on sidewalk lampposts for getting five-minute naps to make up for no nighttime sleep or vending machines where one can get a caffeine jolt pumped directly to the basal ganglia area of the brain. The future cell phone will be just one *neurochip* connected to cochlear nerves and another to the tongue and jaw. I'd just whisper, "Taxi, please," whenever I'd wish to hail one of those levitating vehicles to take me where I'd wish to go. Addiction to social media feeds will be far worse than what it appears to be now. If we are not vigilant, facts will be tethered to a noisy world twenty-four-hour news cycle fed directly from feedback loops fashioned by competing media companies manipulating mixed commercial and political biases. Imagine the future when the brain is plugged to apps for short-term memory when it is needed or when hologram advertisements will appear on street corners; there might be a three-dimensional image of an actor posing as a doctor advising people to tell their doctors to prescribe a weight-loss injection X that can be gotten immediately at an automated medical kiosk around the corner. You will not need to know the time of day; the *neurochips* within you that will know your daily schedule will warn you of each upcoming event of the day at the appropriate hour to give you a chance to get ready.

---

We live under the instinctive impression that there is a one-way movement of time. Time seems to go forward, never stopping, and never reversing. A cycle in time is very different from the reversal in time's direction. Time goes in any direction in fiction, backward, forward, repeated loops. It can tweak the past to alter the future. The 1993 comedy film *Groundhog Day* presents us with Phil Connors, played by Bill Murray, learning to correct his gross faults through a spiraling time-loop of repetitions of a single day. The 1946 film noir *Repeat Performance* is the story of Sheila Page, played by Joan Leslie. Just before New Year's Eve 1947, she shot and killed her husband. Suddenly it is New Year's Eve 1946, giving Sheila a chance to repeat the year with an opportunity to change the course of the previous year. Film and fiction can always reverse and change time's direction. In real life, it's not so easy, if at all possible.

There have always been questions about whether time's direction could be reversed. Any answers would depend not so much on gravity as on entropy (the universe's random tendency toward disorder) or the second law of thermodynamics. That law not only asserts that converting one form of energy to another always loses some energy to heat but also tells us something about the flow of heat in one direction or, as some people like to think of it, the direction of the order of the universe. The typical example is that an ice cube will melt in a warm room because the heat of the room will flow in a direction to warm the ice. Heat flows in the direction from warm to cold. But whoa, what do we mean when we say that heat flows? Heat is just energy. How does it flow? Any particle of matter with a temperature above absolute zero has some heat energy, heat having something to do with molecular vibrations. Hot molecules vibrate more than cold ones, but heat flow is not so simple as to say that temperature is just the rate of molecular vibrations, for some molecules are likely to vibrate faster than others. By that, all such matter seeks energy equilibrium with its neighbors. Too bad that rarely happens with the contentious politics of humans. In any disequilibrium between neighbors there is an imbalance of energy that tends to be dispelled as heat, which suggests that there is already an intrinsic flow of energy in the universe that dictates a preassigned direction to order. Humpty-Dumpty falls, and that's it for him. Entropy in a closed thermodynamic system is the degree of disequilibrium in the system. This sounds vague because it is vague. There is an actual definition that gives an actual number connected to the proportion of molecular configurations in a system, but a detailed definition of that number here would confuse the issue of time's direction of flow. All we need to know here is that entropy tends to increase in any system that is exposed to neighboring systems, such as, for instance, the ice cube in a warm room.

Slow down time? Sure. Any proton can do that. Get a proton in a particle accelerator up to a speed of 99.9999 percent of the speed of light for a year. The proton, if it had a mind, would come back thinking that no time had passed. But reverse time? Not going to happen, even for that proton weighing less than $10^{-25}$ the weight of a human, though there are theoretical time reversals that are possible since they fall just a wee bit short of being impossible. In consideration of that minuscule possibility, there is a theoretical way to go backward in time. I'm referring to a 1988 paper by the Cal Tech astrophysicist Kip Thorne and two of his graduate students in the *Physical Review Letters* that give mathematical instructions for how to do it.[8] Since space—by Einstein's theory—is a very large, curved manifold, time and light (especially light) take a while to catch up with faraway regions of the universe. Wormholes are shortcuts, short

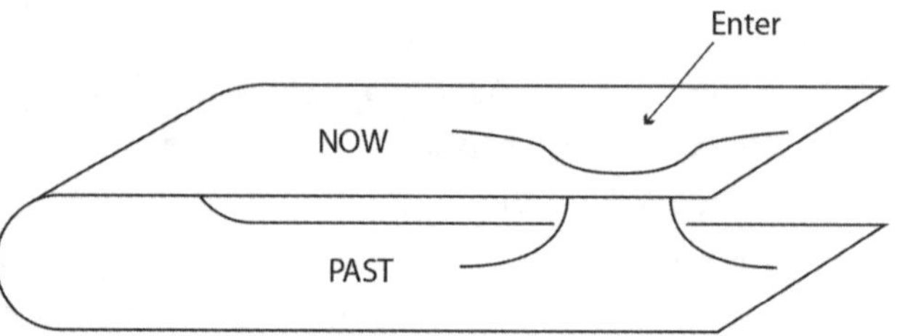


Wormhole for intergalactic traveling

"handles" in the manifold of space connecting faraway places. The illustration here gives a schematic diagram of a wormhole shortcut. In effect it is a quantum electrodynamical time machine allowing high-energy elementary particles such as photons and gamma rays to pass through wormholes for a shortcut to the past. Humans, however, are massive in comparison to photons, so they might have a problem entering the "now" and coming out in the "past" and not ending up as a soup of mixed molecules having nothing to do with being an animal. Could humans use wormholes for intergalactic travel with return-trip tickets? Thorne says, *yes,* theoretically.

But this doesn't mean that humans of relatively massive size can take a short peek at the past and return. It doesn't mean that we can go back and alter history. To time travel, a human body would have to move very fast in a race against time, beat it to the past in a leapfrog of time itself.

Can a time machine be built? That depends on what is meant by an actual machine. The word implies a contrivance that uses energy to perform a task, a device that is not necessarily made of earthly material. So, a time machine could be some idea for framing an almost, but not quite, impossible scheme for advancing or retreating time. The Time Traveler in H. G. Wells's *Time Machine* built a chair made of ivory and some crystalline substance with levers and dials, a metaphorical representation of something that can speed up time, slow it down to a crawl, or even get it to go backward. He tells us that his maiden voyage of time traveling was particularly unpleasant.

> As I put on pace, night followed day like the flapping of a black wing. The dim suggestion of the laboratory seemed presently to

> fall away from me, and I saw the sun hopping swiftly across the sky, leaping it every minute, and every minute marking a day. I supposed the laboratory had been destroyed and I had come into the open air. I had a dim impression of scaffolding, but I was already going too fast to be conscious of any moving things.[9]

From this we see that the Time Traveler did not leapfrog with time; his time traveled along, outpacing the present to advance faster than his surroundings. Eventually everything would catch up. In almost every scheme of physics that I know, a time traveler does the same. In riding world-line loops that cross the past, the Time Traveler still moves along a continuum of forward time. Finding wormholes in one universe to slip into another is the only path that jumps through a shortcut to the future or past. But even in that case, the route is by a loop meeting of times, not a jump.

One of the best general books on the physics of time travel is J. Richard Gott's *Time Travel in Einstein's Universe.*[10] Gott gives readers a catalog of fantastical schemes that are theoretically feasible. All time travel, including Gott's theoretically and arguably possible schemes, assumes that there is a past, present, and future out there somewhere with simultaneous worlds, alternative universes that are moving through time in a manner that is out of syncopation with our own. If some inventive time traveler were to jump back in time from our world to an alternative universe to kill Hitler, he or she might succeed, but that killing will have no effect on our universe. Still, I'm tickled by the thought of some virtual invisible visitor shadowing me.

The wonderful thing about physics and its mathematical formulas is that you can use theory to present some astounding results that are possible without being feasible. Create a model of the real world that involves the symbol $t$ for time, and work that equation past the limits of real world feasibility; then say, *Look at what we can do with time.* After all, equations that are based on common-assumption truth principles don't lie.

So, don't expect to travel back in time yourself to alter your present life. To get to the past from where you are now would take some heavily intelligent and colossally fast maneuvering through wormholes and black holes, using a vast amount of galactic power that does not disturb our galaxy. If by some amazing feat you did manage to jump backward by a few minutes, you would be only a participant observer, powerless to alter the life and time you came from.

But that's relativity theory, with gravitation and all of time's different reference frames. Quantum theory then follows to confuse the issue by bringing up the question of whether time exists at all.

PART

# THE COGNITIVE SENSES

Let one sit with closed eyes and, abstracting entirely from the outer world, attend exclusively to the passage of time, like one who wakes, as the poet says, "to hear time flowing in the middle of the night, and all things moving to a day of doom." There seems under such circumstances as these no variety in the material content of our thought, and what we notice appears, if anything, to be the pure series of durations budding, as it were, and growing beneath our indrawn gaze.
—*William James,* Principles of Psychology

# 12

## THE BIG QUESTION (THE SENSE AND PLACE OF TIME)

Time is not just a *t* in a mathematical physics equation. It is something more in the mind and human spirit than in the workings of cosmic dust. There is, however, a fundamental disagreement between the physicists' notion of what time is and the standard everyday understanding, what we might call the human perception of time, some inner rhythmic temporal imprint that supports Saint Augustine's smart answer to his famous question.

Zoom in . . . Spot the second in the timepiece of mortal lifespan. What is it? Time is trapped in a somewhat circular enigmatic meaning because it uses time-dependent words such as *duration, intervals, motion,* and *change*. Those words hover over undergrowth of connected concepts. Speak of one and we are inevitably led to the others through convolutedly entwined connections. Those words are inseparable in relationship. Though they properly belong together, we commonly speak of them separately. Time seems to always slip into that realm of thought that blocks satisfaction. *Quid est ergo tempus?* (What, then, is time?), Augustine asked, and then answered, "I know well enough what it is, provided that nobody asks me; but if I am asked what it is and try to explain, I am baffled."[1] Immanuel Kant aimed to answer: "Time is nothing but the form of the internal sense, that is, of our intuition of ourselves, and of our internal state." What is that internal sense that Kant is talking about? Space, he understands, has an internal geometric picture. He is suggesting that time, like space, also has an internal geometric picture as a one-dimensional image, an image that comes to us before all our living experiences. Kant considered that internal picture: "We cannot think a line without drawing it in thought; we cannot think a circle without describing it; we cannot represent, at all, the three dimensions of space, without placing, from the same point, three lines perpendicularly on each other; nay, we cannot even represent time, except by attending, during our drawing a straight line (which is meant to be the external figurative representation of time)."[2]

That was written in 1781. He was wrong, of course. But his understanding of time's connection with space is certainly forceful and, perhaps, foresighted. He could not have known that a bit more than two centuries later, physicists would make the old connection between space and time more understandable, and yet more confusing.

We live in the present moment and look toward the future, so we might wonder if there is some boundary to the present that differentiates it from the future. The problem is that the present just happened and whatever just happened has already passed. We have moved ahead imperceptibly in time. That which was once the future is suddenly the present, then just as suddenly the past, to remain the history of everything that once happened. Through our language, we have a vague impression that time is something like a river fed by tributaries of the past that flow from now toward a basin of all future possibilities. Unlike the White Queen in *Through the Looking-Glass,* who tells Alice in a careless tone that she remembers things that happened the week after next, we know that if it's a Monday, Tuesday is set to follow.

Language, in part, is responsible for the confusion. The word *time,* when used colloquially, almost always refers to day-to-day, hour-to-hour time, the time displayed on our wall clocks, phones, and kitchen appliances. The broader use of the word *time,* however, comes from just a single root that shoots sprigs of various imports. I'm reminded of the mistaken claim and persistent myth that Inuit have four hundred words for snow.[3] Even for Eskimos, snow is just a root that needs a qualifier to mean the downy white flakes of frozen crystallized water or the piles of packed snow on the ground measured in inches or the drifts of snow or sleet or the slushy, dirty snow at the edge of roads or anything else that remotely resembles the image of that white stuff that falls from the sky. Like *snow,* the word *time* is used with a generosity of context qualifications that are not limited to solar or sidereal time, relativistic time, biological time, mindful time, or temporal states. It is a word of ample meanings, so perhaps distinct notions of time should have distinct words.

Yet there is that single root that seems to lead to all meanings of time. Whenever we talk about time, whenever we try to dissect the meaning of time, we habitually end up talking about space and change. Time, the principal measure of all change, seizes the notion of space at its core meaning—"to determine a spatial dimension." We live in that spatial dimension and thereby consciously experience all measurements as if they must be marked by some spatial metric such as length or size.

---

One might think time is simply a measure of change in space. If nothing changes, neither does time. That is precisely why it seems as if the hands of a watched clock do not move or the water in a watched pot does not boil. Where are your thoughts then? Looking, waiting for change, and paying too much attention to time. It—whatever *it* is—or him, as the Mad Hatter suggests, very much likes that attention.

It is that attention that grabs us and carries us along the day as mindful time advances though individual perceptions. I know when to have breakfast and know approximately how long it will take to get to work. It is, in Jorge Luis Borges's words, "the substance of which" he is. "Time," Borges writes in his essay "A New Refutation of Time," "is a river that carries me away, but I am the river; it is a tiger that destroys me, but I am the tiger; it is a fire that consumes me, but I am the fire."[4] That's far different from knowing the speed of light in a vacuum or when the next eclipse of the sun will be.

That's why so many notable philosophers have struggled with the reconciliation of subjective and objective realizations of time, wrestling with the fact that human perception of time will always be far from anything that can have a precise measurement. The twentieth-century French philosopher Henri Bergson thought that the living experience of a real duration of time (*durée réelle*) is very different from any time structure defined by science. Time for Bergson was mindful, different from the kind of time that works the clocks. As with love, suffering, the sound of falling trees in a forest, or the consequences of passages in an unopened book, it does not simply exist on its own without the mind that perceives it. At the heart of time is simultaneity. The clock moves to noon, and something having nothing to do with the movement of the clock happens some distance away. But that simultaneity—or at least the movement of a clock's dial—has little to do with cognition and memory representing temporal judgment.

Bergson's belief follows that of the eighteenth-century Irish philosopher George Berkeley, known as Bishop Berkeley, who wrote that "neither our thoughts, nor passions, nor ideas formed by the imagination, exist without the mind." For Berkeley, time is an imprint in the mind initiated by imagination, a continuum of ideas, one after another, a rhythm of human consciousness.[5]

Perhaps time is just that rhythm, a beating awareness of life moving on. The causes and nature of human consciousness that had been persistent topics of debate since Plato's time were energetically revived when René Descartes questioned his own existence. At the end of the nineteenth century, the psychologist William James applied continuity of time to his investigations of what he called *the stream of consciousness*, arguing that it is impossible to stop any thought for introspection before it reaches a conclusion. If, with some luck, the thinker is "nimble enough to catch it, it ceases forthwith to be itself." It seems that a conscious thought evaporates before it can be examined, like "a snowflake crystal caught in a warm hand." Any attempt to freeze the continuous stream of a human's conscious thoughts is as pointless as stopping "a spinning top to catch its motion, or trying to turn up the gas quickly enough to see how the darkness looks." These are "as unfair as Zeno's treatment of the advocates of motion, when asking them to point out in what place an arrow is when it moves."[6]

For James time and space are interconnected wholes that could never be thought of separately. Any conscious thought of time brings with it a notion that time is moving—a spatial notion akin to Minkowski's spacetime.[7] That notion comes from living in time, and yet we don't feel it by any conscious internal sense. Perhaps that is because unlike time, consciousness itself is discontinuous, sporadic, often interrupted, and resuming. It is disrupted by sleep and dreams; yet it *seems* so continuous, like the motion of a zoetrope that is nothing but the illusion of continuity. We may not be able to tell whether conscious thought is continuous, but we do know that the complex bundles of signals perpetually collected from all human senses are tidily synchronized and recorded to form what we call *consciousness*. And so, we shouldn't be surprised to learn that the signals of time are neatly packaged within those bundles.

Christian Marclay's *Clock* is possibly the only long movie that can be watched with total background consciousness to time's passing in a continuous collage of bundles. It is a masterpiece of editing over eight thousand movie clips into a twenty-four-hour collage of overlapping clips of sounds and scenes that keep the viewer gripped in a continuous awareness of time for every one of the 1,440 passing minutes. By contrast, Andy Warhol's *Sleep* lasts 5 hours and 21 minutes, making time freeze for at least the first 45 minutes of watching a man's abdomen rising and falling.

Questioning time has almost always followed questions of whether time exists outside our own thoughts about it and whether time exists beyond the present moment, that moment that is gone the instant it arrives. Is it real? Is it illusionary? Is it manufactured? Is it integral with our consciousness? Is it relative or absolute?

Here the present moment is meant to be that strict invisible point that theoretically divides the past and future, that moment when, as James suggests, we might foolishly try to turn on a light to catch a glimpse of the darkness. Such a point exists in theory, and arguably it actually exists, yet the only real way to think of it is by loosening the strictness and envisioning an interval.[8] That imagined interval, however, is a part of space. So here again, we find our vision of time represented as space.

---

Is there some special faculty that permits us to be aware of time as it passes? Perhaps what actually happens when we feel the passing of time is that we are relating time to something real, an event, an immediate experience, a history, a story being told, a memory. Kant maintained that experiences endure in memory and that the objectivity of that memory endures and becomes the underlying substance of thought. That underlying substance of thought is what builds mental time. For Kant, the "something" that endures in memory coming from a person's experience is analogous to the "something" in finger memory of a pianist who from time to time is not even paying attention to the notes on the score. Of course, fingers by themselves have no actual memory, no mind of their own to think about what the next note should be, and when and how long it should be played. It is the brain that knows what note comes next from a memory of the music that manipulates the muscles. Finger movements take their orders from someplace in the motor cortex with instructions from someplace in the brain campus that recalls the experience of the music being played, along with expressive timing and tone.

William James called it *internal perception,* but for him it was simply the memory of past time. He asked, "But how do these things get *their* past-ness?" Then he brightly hypothesized, "Our feelings are not thus contracted, and our consciousness never shrinks to the dimensions of a glow-worm spark. *The knowledge of some other part of the stream, past or future, near or remote, is always mixed in with our knowledge of the present thing.*" His glow-worm reference suggests that the insect uses its bioluminescence to shine only on its immediate area, keeping its past positions in darkness. And unlike the glow-worm's glow, conscious objects of the past linger in the present and fade as "now" objects come into consciousness. It is a nineteenth-century descriptive linkage between consciousness and internal perception. James's narrative model suggests that consciousness is a continuous stream, rather than a succession of events, in which the past lingers just long enough in the present to give a smooth feeling of event transition. James puts it this way: "If the present thought is of A B C D E F G, the next one will be of B C D E F G H, and the one after that of C D E F G H I—the lingerings of the past dropping successively away, and the incomings of the future making up the loss." In other words, consciousness itself is a sequence of impressions and feelings that collect and dissipate in time. Some of those feelings coexist in just enough duration to give the illusion that consciousness is a stream rather than a discrete sequential collection of independent thoughts; one follows another, as one grows fainter in memory and another stronger. All this happens in the present when two feelings delicately coincide, "a sub-feeling that goes and a sub-feeling that comes." One is evoked, another is presumed. One thaws as another strengthens. All this happens in a rapid cadence of unfathomably short instants that move like a stream.[9]

We could think of the internal perception of the present, this continuous movement from a sub-feeling to a sub-feeling, as the passing of real time. Such a model rightly puts consciousness as the conveyor of the present, a present that moves forward. "In short," James tells us, "the practically cognized present is no knife-edge, but a saddle-back, with a certain breadth of its own on which we sit perched, and from which we look in two directions into time. The unit of composition of our perception of time is a duration with a bow and a stern as it were—a rearward- and a forward-looking end."[10]

That sub-feeling that goes to a sub-feeling that comes is conveyed in the idea of a zoetrope, a nineteenth-century parlor room toy that gives the illusion of motion, only in this associated instance it is giving the illusion of the consciousness of time. The zoetrope is nothing more than a spinning cylindrical drum with slits and a sequence of still images of, say, a horse with feet and body in running position. Each image is very much like the next, except for a slight difference in anatomical position. When the succession of discrete images is viewed through the slits of the turning drum, the viewer sees the images fused into a dynamically moving picture—a horse in motion.

Just as still images in rapid succession are interpreted as real continuous motion, a rapid succession of nows, each much like the next, except for a very slight change, is interpreted as a continuous flow of time. The celebrated nineteenth-century physicist Hermann von Helmholtz, in writing one of the great contributions to medicine, his *Handbook of Physiological Optics,* suggested that the zoetrope gives us the illusion of motion because the eye holds one image just long enough for the next to take over. Something of the sort does happen in the retina; look at a black spot on a white background for a few seconds and then turn away. The black spot will linger for a few seconds more. This is even more pronounced when we see a spot of bright light in a dark room long after the spot of light is extinguished. The spot is temporarily *burned* onto the photosensitive retina. But we now know that the coordination of discrete visual images—real sight—takes place in the visual cortex, not in the eyes, so the question remains: How is it that a rapid succession of still images is construed as a moving picture seamlessly flowing in time?

---

Knowing time, even just one hour, is never brightly clear in the mind. James believed that we never get the full cognition of any precision of time, yet we do have a pretty good perception of space when we look out a window at a panoramic scene. What happens when we look away from that panoramic view in an attempt to isolate time as a vacant moment? No sight, no sound, and no outer world, in what James called an *indrawn gaze*.[11]

James believed that the indrawn gaze of time itself is not devoid of sense content. When you close your eyes, there is always a bit of light, luminosity through the thinness of eyelids, and sound, through ceaseless body murmurs, the heartbeat, the breathing, the rhythmical expansion and contraction of the chest, and of course the faint hushes coming from the inner ear that we almost always ignore. So we are never really as indrawn as we think. Wilhelm Wundt, who is often called the father of psychology, called such a state "the twilight of our general consciousness."[12] It is in some way a virtually vacant consciousness. Yet it is a moment when we do become aware of passing time in Wundt's twilight of consciousness.

Let's not take James's or Wundt's word for it. Close your eyes and attempt to achieve that indrawn gaze. If you have some success at emptying the content of thought, your indrawn gaze will suck into the void some words and images in an on-and-off regular awareness of the task. That's inevitable. Your mind can never be totally empty, but what will be achieved is a sense of the rhythmical process of an almost empty consciousness, and there, in the dark recesses, lurks a sense of the flow of time. It's as if the mind and body rhythms work together to clock our live presence in the background of our noisy routine consciousness. Eliminate the noise of routine consciousness, and you have the feeling of watching time flow in beats of some kind of unspecified discrete unit—*now, now, now*—with the illusion that it is continuous.

Everyone knows what time is, in a sense. Even so, with all the diggings into the depths of time, it all comes back to Augustine's musing. We know what time is, as long as no one asks us what it is. Going further gets us into metaphysical snags that confuse our personal feelings and go far beyond the scope of comfortable understanding. We have the integration of scattered thoughts of so many philosophers, scientists, and psychologists that could help us know time better. But to know time better, we must continue to know how we got here and how time brought us here, here to this immediate present, whatever that is.

That immediate present is the moment of living, the moment of thought, the moment of action. When hungry, your mind can advance to think about what you will soon eat. Your in-the-moment sense is in anticipation of what will happen next, with a temporal understanding of where you are now. It is your consciousness that is simply making its way toward the future while pushing the present toward the past. The present is a *you are here* mark, but that mark is a moving target, always updating itself, always accepting new material passed along from the future. That is why we think that time flows. In English, we read from left to right, we see the words on the line coming long before we process any comprehension. Those moments of comprehension are the *nows* of our reading. The open future keeps coming into the present without blanks, dumping the present into the collected volumes of the past. The circular clock hand points to the present moment (or as present as any moment can be). We know that clockwise arcs mark future moments and that counterclockwise arcs mark what just passed. So there is a *flow* downstream from future to past. We also know that when we scramble an egg, we have moved the existence of that egg to whatever it has become in time. And we can detect the awkward tense incompatibility of the previous sentence as deliberate and unavoidable.

Time fills the timeless cabinet of paradoxical curiosities: the now becomes the past in a flash. The now expires like a soap bubble at any attempt to consciously grab it while the future slides into now's void. It seems to measure all things, yet it itself defies measure and all attempts to be measured. It "advances like the slowest tide, but retreats like the swiftest torrent. It gives wings of lightning to pleasure, but feet of lead to pain. It lends expectation a curb, but enjoyment a spur."[13] All attempts to get it seem to be like grabbing a fog. An annoying illusion, yes. But not totally futile. There is something there to grab.

Can we ever know what time is
for a nine-year-old who daydreams
her tenth birthday a week away?
The Olympic slalom racer,
imagining her course, knows it
all too well while buckling her boots,
as long-haul truckers, months from home,
do see endless roads of boredom,
and convicts in solitary
count their never-ending days.
*—JM*

# INTERLUDE: TIME IS WITH ME

Every now and then I think that time is there with me, and yet nowhere, that it is a fabrication built in my head, as if something real that has to be obeyed, when all along it is just an illusion made from being human in a world that must cope with coordinated meetings of things in places. Of course, I know that it is more than that. Without time, the world would be muddled and messy, if not totally chaotic; with time, humanity is at least capable of working as a harmonious whole. It gives us form, number, and measurements to further our understanding of cause; we have beats of poetry, sounds and rhythms of music, openings of daffodil buds, longings for opportunity, expectations, and memories of accomplishments. And yet I sometimes think that time is a fabrication no different from the alphabet that permits us to organize stories and information. Can anyone say that the alphabet is real, and not created? Did it exist in prehistory when the vocal cords of modern humans uttered the first vowels? Just as the alphabet is an organizer of written word sounds, so time is an organizer of human situations and circumstances in the communal organic lifespan. Perhaps time is just a made-on-earth artifact that enables us to know that life is brief and that there is so much to make of it. I know that time is the spine of history, the cataloger of existences and survivals, and our peephole of hope. And yet that thinking, heavily influenced by Einstein's division of past, present, and future as "a stubbornly persistent illusion," dithers the more I try to think about what time is. Why? Because, until recently, I had always thought of time in terms of what it does. Don't we all? But, like Achilles's attempt to catch the tortoise, I could never catch up with a winning understanding of what time actually is. I'm glad to say that I am getting close. Maybe that's all anyone can hope for.

# 13

## WHERE DID IT GO? (ACCELERATION OF TIME IN AGING)

> Wherever anything lives, there is, open somewhere, a register in which time is being inscribed.
> —*Henri Bergson,* Creative Evolution

There is a certain age of life when, on occasion, conversations with friends tend to center on the perception that time is speeding up. It's the age when life seems to pass at a forever-increasing pace. Unlike a six-year-old, who daydreams her seventh birthday that is a tediously long week away, a sixty-year-old conjures that same week as a flash. Without some sporadic time-marking adversities, annoyances, or pleasures to shake up routine, a seventy-year-old recalls times between breakfasts and dinners as vanished flickers of life.

As we age, our metabolism and body clocks run slower, even though the sun attempts to readjust those clocks every day. Older people generally move more slowly than younger people do, not because they cannot move fast—they often can. They are not usually aware that their movements are slower. When body performances (balance, reaction timings, vision, hearing) slow down, the relative motion of time appears to slow down as well, so one does not notice that time has advanced at a different pace than when one was younger. Watch people over seventy and young folks move through the aisles of a supermarket. For many of those older folks, time sense is adjusted to what their bodies, especially feet, are signaling to temporal components of their brains. Some feel that their lives are passing faster and faster with age, yet their minute-to-minute clocks seem to run more and more slowly. Older folks generally are subconsciously and consciously aware that they have more time, though the percentage of time remaining alive is rapidly diminishing. Retirement has significant effects on how much time people have on hand, but their movements are not dictated by their free time. With free time comes an attitude that encourages a slowing down of motor activity. That slower pace contributes to adjustments of the internal clock that continuously tell the body how fast or how slowly to move. Internal clocks tell those worn springs, gears, levers, and pistons of their bodies that there is no rush to get anywhere. They confront any psychological ponderings they might have about how much time is left before death.

Pierre Lecomte du Noüy was chief of the Division of Biophysics at the Pasteur Institute in Paris in the 1930s when he studied the repairs of wounds received in the battlefields during World War I. He established what seemed to be a universal relation between the speeds of wound healing and age. His work was inspired by Alexis Carrel's work on wounds. Carrel was a French biologist and Nobel laureate in physiology for his work on techniques in vascular surgery. He suspected that the cicatrization of wounds followed some geometric law but couldn't completely prove it without du Noüy's mathematical help. Du Noüy was able to express that rate by a rule that resembles the function that fits closely with the relation between rate of cicatrization and age. We know now that the rate of healing the body after any kind of major surgery slows with age.

Du Noüy went through several chemically based experiments to confirm the relation between the rate of cicatrization and the psychological impression of time flow. He found that a twenty-centimeter (almost eight-inch) wound of a forty-year-old man will cicatrize at a rate close to twice as slow as that same-size wound on a man of twenty and that such a wound on a man of sixty will cicatrize five times more slowly than that same-size wound on a child of ten.[1]

If the human body repairs itself faster when young, one could imagine concluding that the rate of time passing affects body and brain operations at the cellular level. Of course, we must accept that a child's view of the world is different from an adult's—that is, the child is constantly embracing far many more new experiences than the adult. Hence, time does not flow evenly between the ages. We are talking about time in the human consciousness as if it is dependent on the cells of the human body. That might seem far-fetched. Du Noüy was not suggesting that cell time (whatever that could mean in the 1930s) was the only factor contributing to our consciousness of time but rather proposing that that physiological time must at least amend our psychological time. His experiments showed a close quantitative correlation between the rate of cicatrization and the psychological impression of time flow. His correlations suggest that cell repair rates, and perhaps other metabolic functions, have some influence on our inner sense of time flow. Therefore, it is possible that our inner senses of time are connected in some ways to cellular activity.

We know enough about the chemistry of the body to know that with age there are increased toxins coming from both the natural deterioration of cell tissue and the cumulative onslaughts of environmental contaminants. I'm not saying that those contaminants are more toxic to organs than they were at any time in human history. They probably are, but the more relevant point is that until our body organs evolve to defend themselves better against those environmental contaminants, the effect is that older folks will accumulate more toxins that embattle tired regenerating cells while they continue an aging life. That is the aging process. With such a thought we can compare the quantitative correlation between the rate of cicatrization and the psychological impression of time flow, as illustrated here, which shows a graph of the rate of cicatrization as a function of age. With age and the more toxic accumulations in the body that are not fought off comes a slower healing rate for wounds, and likely for other ailments. Also with age comes the notion that the apparent length of a year is nothing more than a subconscious understanding of proportion of life gone by, or a subconscious compounding of memories. Perhaps, however, those psychological notions are confounded and correlated with the physiological aging of cells, something very debatable without knowing a cause.

Whenever I find myself mulling over the problem of perceived time acceleration with age, I enter a zone of puzzles mired in the meaning of perception itself vis-à-vis awareness that there is a thing that causes time. *Perception* is as tricky a word as *time*. I know what it means by definition:

the ability to become aware of something that is happening or of something that could happen. I can perceive an event that happened in the past and events that will happen in the future. But perception is not the same as memory; it's broader. I can perceive things that never truly happened and perceive things that I expect will never happen but could.

When a thing truly does happen, we tend to forward-telescope its time of happening. *Telescoping bias* is the term cognitive psychologists use to tell us that we tend to recall recent events as happening farther back in time than they did. Likewise, we tend to view remote past events as being more recent than they were. Consider an important event in your life that you have not thought about in a long time: the marriage of a friend or your memorable first kiss. When did it occur? Chances are your memory will fool with the date, sometimes off by years, playing with the sense that time speeds up as we age.

Many theories try to explain the phenomenon of perception of time in aging. The most popular but wrong one is that as age increases, a current year of life is a decreasing proportion of a whole life. It is an old idea attributed by William James to the nineteenth-century French philosopher Paul Janet. James tells us that it "can hardly be said to diminish the mystery. . . . A child of 10 feels a year as 1/10 of his whole life—a man of 50 as 1/50, the whole life meanwhile apparently preserving a constant length."[2]

It's a linear model of human temporal impressions listing percentages of a year in the life of a person *N* years old. A year in the life of a ten-year-old adds 10 percent. Two years would have to pass in the life of a twenty-year-old to add that same 10 percent. Continuing this idea, we see that it would take *N* years for a *10N*-year-old person to have 10 percent of his or her life seem to pass by. It gives a mathematical line that appears to be close to what older folks feel is true, but the cause for that feeling is far more complex.

This proportion-of-life theory assumes that each passing month or year is less weighty relative to a whole life. But it ignores the small, significant moments that can occur at any time interval in a person's life. The impression of foreshortening of years as time goes by is more complex. It depends on new and routine experiences, on memories, on travel, and, very strongly, on anxieties. As the years pass, each year automatically becomes more routine, and the repetitiveness seems to lump the days, weeks, and years into narrow bundles that are counted as small envelopes of time.

The linear model is the typical answer to the question of why time seems to foreshorten as time goes by. One year in the life of a person $x$ years old is less than for a person $y$ years old when $y < x$. Of course, that is true mathematically. It does—and does not—square up with reason. Neurological studies tell us that the dopamine system, with its receptors and transporters, weakens with age.[3] Pharmacological studies link dopamine systems to the internal clock and temporal memory, duration discrimination, and attention mechanisms.[4] Psychologists have used small-sample techniques to study whether age plays a role in time counting. Subjects were asked to count seconds for three minutes by saying, "1–1,000, 2–1,000, . . ." The younger group, aged between nineteen and twenty-four, counted the seconds away almost perfectly. The older group, ranging in age between sixty and eighty, were off by forty seconds. Not far off, but if we were to extend the counting for, say, one hour, it would amount to more than thirteen minutes. The older subjects were not retirees; they were working people on tight schedules, so the implication is not that their senses were affected by boredom but rather that their brains' internal clocks run slower, giving the sense that life is speeding up. It might be a result of the weakening of the dopamine system.[5]

The brain controls reactions and estimates durations. Babies cry when things don't come at expected times, when a mobile over their bed stops turning too soon, or when food does not arrive as fast as before. Sure, they experience a discomfort in the belly, but they are also shaping expectations. Young people are very aware of time; commonly, they think about it when they are bored more than older people do. They tend to watch clocks more than older people do. They check on how long it takes to do something. And this awareness improves all through childhood as long as the prefrontal cortex is maturing.

Experimental psychology has been studying time perception for more than a century with great success in many ways.[6] Some arguments tell us that animals share some time instincts with humans, that time is centralized in the same area of the brain as that which processes such rewards. Other arguments connect time management to areas of the brain that

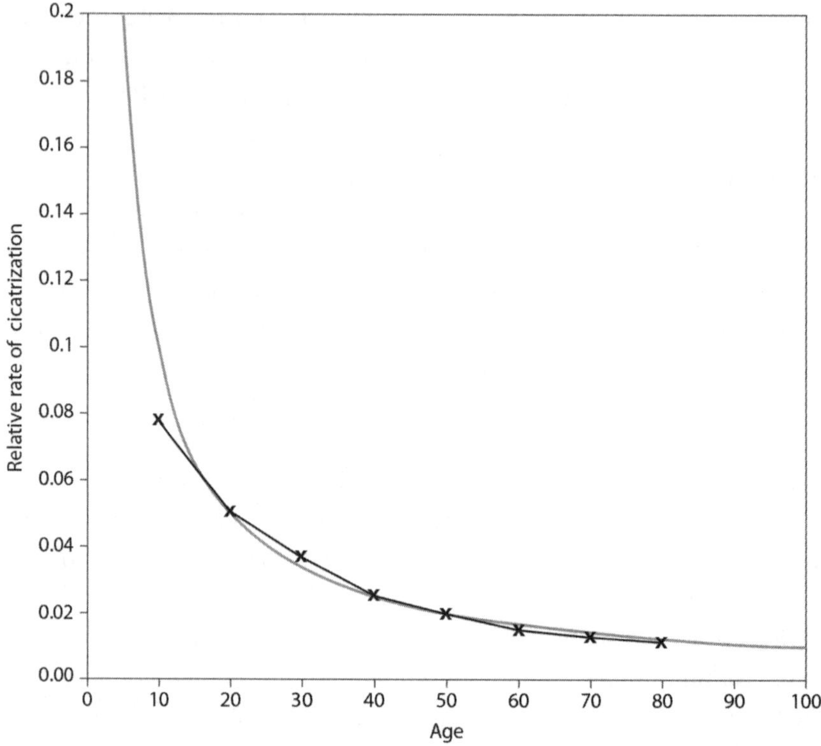


Relative rate of cicatrization and appreciation of the value of time as a function of age

process anxieties and fears. There are arguments suggesting that many of the regular rhythms of the body, such as heartbeat, respiration, pain, temperature, and emotion, contribute a gestalt for time. It could be that time perception is somewhat linked to the consciousness or unconsciousness of body rhythms and heartbeat, giving the heartbeat more credit than it deserves.[7] The problem for all theories and arguments is that there are many factors, some apparent and some hidden, involving brain mechanisms designed for internal cell timings that contribute to our conscious sense of time.

Illustrated here is a rather simple hyperbolic curve representing Paul Janet's psychochronometry model of human temporal impressions of time coinciding with a curve representing physiological time of wound healing. It is just a resemblance between the two, not a cause. We know that many natural phenomena are modeled by hyperbolic curves. There could also be other hidden processes, neither psychological nor physiological, that conflagrate the presumption—a welcome thought that something more than psychology might affect our sense of time. Of course, the connection could be, like most connections, a simple coincidence. Or should we view this as no real surprise, since many chemical processes follow similar geometric progressions affected by the physic-chemistry of cell life?

---

Returning to that old question of why we think time is speeding up with age, consider the familiar idea that a person envisions his or her life in terms of the percentage of the years that have passed. A year to a five-year-old girl would be one fifth of her life (20 percent), while that same year to a twenty-five-year-old would seem like just one twenty-fifth of her life (4 percent). If we examine this relation further, where $x$ represents the age of a person and $y$ represents one year's portion of his or her life, the relation is simply $y = 1/x$.

Age definitely intervenes with our impression of time's passing, but not in the way we might expect. The theory of time awareness as proportions of age, both conscious and subconscious, is rational, but that kind of

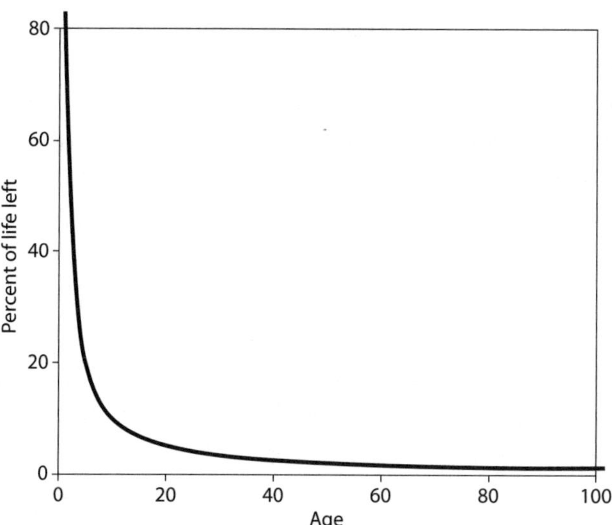


Percentage of life left as a function of age

sentience is built on memory of an entire life, albeit a blurry sketch of a life. Du Noüy put it this way in a wonderful metaphor:

> [The mechanism of memory] is somewhat as if we were in a theatre watching the play of actors—representing our internal perceptions projected outside of us—and as if the scenery were being shifted always in the same direction, unwinding itself on the right and being rolled up on the left. From a mechanical point of view it is easy to conceive that the more canvas is wound the larger will be the diameter of the roll and the greater will be the velocity.[8]

That unwinding-winding makes rational sense, yet it may simply be that with each passing year of life, our daily experiences are more repetitive and humdrum, with fewer new ones of any lasting impact. We don't pay much attention to routine experiences but do pay significant attention to salient ones, such as a first kiss, a first car, a high school prom date, or a fall that chips a front tooth. They are landmarks on trails of lives. But salient experiences like those first kisses, cars, or graduations get less frequent with age. Possibly that's one reason why we think time speeds up with age, though, by far, it is not the only reason.

We humans are far too complex to have our time senses guided solely by a single series of common actions. The causes are not so simple. Whatever steers the time's sense of accelerated passing must come from a multifarious hierarchy of intricately related inner signals that interact with the body's understanding of how the life we lead convolutes with time. One cause in the hierarchy is a person's activity. The active life gives the impression that time is moving slowly, contradictory with the notion that boredom makes the clock run excruciatingly slowly. That activity is intimately connected with body time, pulse, heartbeat, body temperature, muscle memory, and brain activity. Time sense is one giant ball of interconnected feedbacks from the way we live.

Thinking about the end of any duration of routine activity is influential. A month's vacation at a spa resort is a block of time with a beginning and an end. One envisions the end as it approaches. At first, there is no routine; rather, there is a somewhat empty future to be filled, a sequence of indefinite experiences. Meals and mealtimes are changed, sleep cycles are changed, and so are attentions to time itself. That block of time, with its preestablished end, plays with the mind. Time seems to pass slowly at first. Then, after a few days, the empty future fills in to create a new routine. As one adjusts to that new routine the lengths of days seem to accelerate toward the end date of that duration. Often there are biophysical changes, blood pressures, sunlight exposures, and drinking practices that complicate matters to confuse time.

Proximity to the end of life is instrumental to the reason why we think time speeds up with age. It is the body's vigilant signal of the coming of that good night, the weary cell's cautionary warning of the impending expiration date, the end date of our longest block of time. We know that day is coming but not when. As we age, we become more and more aware of the "when" as our aging cells become increasingly lax and confused about coordinated functions, such as when in the circadian cadence to turn protein building on and off. Also, we learn of more and more deaths to find life more and more ephemeral than we once thought. The mind, then, attempts to snub forever-increasing background dins that whisper whims about the number of years left. Eventually, we all come to an age when the cells of our body tell us that the years seem to be going by fast and that next year will seem shorter than this one.

Thankfully, there are many moments of pleasure, moments when we bite into something delicious, moments when we are in the flow of an energized focus, perhaps enjoying a spring walk in a garden, perhaps rhapsodically relishing a meaningful piece of music, perhaps being absorbed in a favorite hobby, or perhaps just in moments of deep intellectual focus. Those moments are often buried and mixed in the timelines of life, but they are the intervals of concentration that move us along with meaning. They come and go throughout our days, jerking, contracting, and dilating time. They are the moments, intervals, and significant durations of times when you feel alive.

## INTERLUDE: THE SLOWEST CLOCK IN THE WORLD

I was fifteen when I got my first summer job. A counterfeit birth certificate proving that I was old enough for working papers permitted me to work at my uncle Jack's silkscreen business in midtown Manhattan. From 8:30 a.m. to 4:30 p.m., I stood in one place like a simple-circuited robot, moving large, ink-wet poster ads for Coca-Cola from a squeegee board to a drying rack, one after another—*extend both hands, grab two corners, lift, turn, slide into rack, release corners, turn back, and repeat.* As if that dehumanizing repetition were not enough mental pain, a gigantic Seth Thomas clock that might have been originally designed for a big-city train terminal was mounted high on the wall directly in front of my drying rack, the slowest-moving clock in the world. Its second hand seemed hardly to move at all, so I found myself always computing a high rate of posters per minute, thinking that the clock was controlled by the front office to pack more work into each day. I lasted just one week. Or was it a year?

Time in that shop stood still for most of every day.

"Why?" I asked myself. "Why did the hands of the clock not budge when I looked at them, and move when I didn't?"

Other innocent questions surfaced, followed by some answers. Why time fools us, sometimes contracting and sometimes dilating. Why it speeds up with age, at the end of a vacation or on beautiful days at the beach. Why it slows down at the beginning of a trip, in an emotional change, or when waiting for a boring lecture to end or for water in a pot to boil. Why it takes forever to drive along a monotonous stretch of highway. Why jetlag happens. Why jetlag doesn't happen. Why prisoners, drug addicts, people with clinical anxieties, dementia, and other cognitive disorders have altered senses of real world time. Why some people have a more precise awareness of time than others. And, in general, why some behaviors so often distort natural temporal thinking.

I still don't have all the answers, but time and I have come to a mutual understanding. I now give it my attention. It pays me back with a headwind against going gently into that good night of my being's end. Time does like attention.

# 14

## FEELING IT (A SENSE OF TIME)

When a man is asleep, he has in a circle round him the chain of the hours, the sequence of the years, the order of the heavenly bodies. Instinctively he consults them when he awakes, and in an instant reads off his own position on the earth's surface and the time that has elapsed during his slumbers.
—*Marcel Proust,* Swann's Way

Two hundred millennia ago, when free-wandering pachyderms or herds of hump-shouldered mammoth roamed the world, was a strange island of time for *Homo sapiens,* who had little free time for thought other than how to survive the hunt while being hunted. In their busy hunting days they lived life in the rhythm of days to nights and nights to days. Seasons were just hints of passing days and years that might have been noticed by clues of the sun's sailing through the sky, perhaps blown by one-directional winds, or by falling like a glowing stone over horizons of hills at the ends of the world. Their sun sailed with no particular regularity, for a day or two, or three, before disappearing for other days under intermittent appearances of clouds that brought morning and afternoon light.

Of course, no one really knows what went on in the minds of anyone living two hundred millennia ago. I've spent several years combing through tomes of anthropology literature searching for any conclusive evidence of protohuman mental keenness, only to come away with my own highly speculative thoughts. So treat what I am about to say here as moderately fanciful conjecture. It seems to me that sometime between Middle Paleolithic and Neolithic times, some hundred millennia ago, someone somewhere looked at the sun, ignored the few days of cloud cover, noticed that daylight came with inflexible regularity, and noted repeating intervals that were each roughly the same. It would have been a simple noting of time. And then it might have taken ten thousand years before someone realized the advantage of using sunlight to construct an interval to measure the past from the present, and then another ten thousand years for someone else to reach a notion of its usefulness and elementary mathematical meaning that could be used for planning the future.

Men had to hunt and to stray far from recognized comfortable safety zones of their families in the warmth of fire. They dealt with ordeals of menacing elements, possibly snow and freezing temperatures, or hunger and pain. Surely there were weaknesses from the bruises of long, fierce hunts and, worse, daily threats from nearby ravenous beasts stealthily looking for supper. Small intervals of time were not distinguished, and neither were large ones. However, in some era very far back, time must have surfaced in the *Homo sapiens* mind as simple recognitions of repetitions, a noticing of seasons, of night and day, so life could have some natural progression through waking intervals of hunting, looking for water, and surviving. And yet those people some tens of thousands of years later seemed to have found bits of leisure time to paint on the walls of caves with sticks of charcoal, carve lunar calendars, and perform ritual events applauding the supernatural owner of the wildlife they hunted or thanking the shamans.[1] It seems that they had instincts for survival and that they organized whatever waking hours were available to hunt, eat, and even keep a bit of luxury time to paint. How could they not have some primitive notion of time? And yet it seems likely that there were humans who had no conscious notion of time.

Perhaps there are, still. Observational and field research into the language of a small indigenous group of people living in the Amazonian rain forest concludes that time sense, the way we know it, is not universally intrinsic to humans. According to research by Christopher Glyn Sinha, distinguished professor of cognitive science at Hunan University, the Amondawa seem to have no structured linguistic conception of time. This fact defies what linguists have long assumed, the *universal mapping hypothesis,* an assumption that all languages have a spatial mapping for structuring temporal relations, alluding to the notion that movement and location are intrinsically tied to the concept of time. To be sure, the lack of a linguistic conception of time does not mean that the Amondawa are a people without a concept of time, since it is likely that the concept predates language. The research team found that the Amondawa have no words for such concepts as "next week" or "next year," contributing to their followup finding that none of the people have an age. In place of age, they go through name-change stages reflecting life divisions within their society. This would imply that time is a dependent of culture and is not universal to human linguistic cognition. But it does not imply that the Amondawa do not talk about events in some chronological order; they certainly do, for instance, order the birth of a child, marriage, and the death of an elder.[2]

And what about that popular belief that the Hopis have no concept of time? It followed a linguistic study done by the American linguist Benjamin Lee Whorf in the early twentieth century. Whorf found no words or grammatical forms relating to past, present, or future, and concluded that a Hopi has "no general notion or intuition of TIME as a smooth flowing continuum in which everything in the universe proceeds at an equal rate, out of a future, through a present, into a past; or in which, to reverse the picture, the observer is being carried in the stream of duration continuously away from the past and into the future."[3] If this seems counterintuitive, it is, because it is likely false.[4] The Hopi have quite a few words and grammatical structures that reflect the use of time in everyday life. Browse a Hopi dictionary and grammar to find past and future tenses.[5] As we found with the Inuit's four hundred words for snow, myths do not retract easily once they reach the surface of mainstay popularity. So, I do wonder how long the counterintuitive Amondawa linguistic probe will hold its account. Context often complements understanding. It is also possible, then, that the Amondawa and Hopi languages do not reflect any of the usual connections between space and time in the same way they connect for us. Humans can maneuver through the pragmatics of life guided by culture without having to articulate experiences marked along some moving timeline.

Animals generally have time senses necessary for survival, with an anticipation of the future as a swollen instant of warning. Of course, we are far from knowing what an animal feels, even though we hope that a dog that wags its tail is happy. Yet we do know, or we think we know, that animals are born with instincts that keep them at least one step ahead of their predators. Those that are hunters must plan their attacks on their prey.

I'm reminded of an experience I had in South Africa, near the border with Botswana. My wife and I were on safari in the Madikwe Game Reserve one afternoon when we came very close to a large pride of lions on a hillside. The lions were crouching, watching three giraffes drinking from a pond at the bottom of the hill, seemingly unaware of the danger. We waited quietly in our open jeep, watching lion eyes—yes, we were that close. We could imagine what the leader was thinking, as if lions could actually think using words: *Should we attack? If so, when?* A lion may be king, but even a king with an army must instinctively be aware of unfavorable happenstances. One kick to a lion's neck by the leg of a giraffe is easily lethal. A lion knows that.

We breathlessly watched the pride. Each lion was motionless, except for nervously rapid blinks of eyelids. The leader was pondering risks. In the end, the giraffes finished drinking, straightening their lanky legs in demonstration of just how tall they really are. Then, with the confidence and grace that such a beautiful animal has, they smartly sauntered away from the pride's temptation. Meanwhile, the pride made its decision: *Better not attack a fully erect tree-eater so tall.* That pride had envisioned a temporal future and planned for it.

We modern humans have far fewer instincts than most other animals because we have a consciousness that advances learning appropriate behavior for survival and therefore no longer prioritizes survival instincts. Whatever instincts we do have are remnants that were beaten down and supplanted by the skills and powers of learning.[6] The American philosopher and cognitive scientist Daniel Dennett tells us, "The brain's task is to guide the body it controls through a world of shifting conditions and sudden surprises, so it must gather information from the world and use it swiftly to 'produce future'—to extract anticipations in order to stay one step ahead of disaster."[7] So our brains are constantly processing incoming temporal information and assembling predictions in advance of any decisions that have to be made in favor of our maximal opportunities for our best chances of survival.

The modern human sense of time is nurtured by experiences in the physical world. We are sometimes conscious of the *now* and sometimes not. Driving to work or to an appointment, those *nows* irk us when we see traffic ahead while the clock on the car dashboard tells us that we will be late. Those conscious instants of *nows,* those present moments, are quickly cataloged in the past as mere memories. The past is fixed to the extent that those memories are not blurred by the interference and confusion of perpetually oncoming new entries. The future comes into the *nows* at rapid speeds, flashing by the realities of the world we live in with contributions from an imagined world. We make plans based on our experiences. They don't always work out, but the conflagration of experience and imagination brings us to intersections of roads taken and not. Unplanned rain can ruin a beach vacation.

Like us, our cave-dwelling ancestors must have built their timelines, moment by moment, pigeonholing their witnessing visions of personal events into an orderly memory of the past—the birth of a child, the kill of an antelope, the death of a mate. Like us, they were archiving their lives somewhere in the centers of their brains. Yet their timelines were unlike ours. Surely, they were well aware of the turning of days, perhaps even somewhat aware of the turning of months from the phases of the moon, and—for those who lived far from the equator—aware of the turning of seasons. But, most likely, their timelines were not chronological in any modern sense, with what must have been blurs or blotches of *nows* and *thens,* with no recall for where in the sequence their memories of events happened. And yet, they were documenting life events for some future purpose.

If we can imagine the caveman hunter spending precious time carving a stone that will later become the point of his spear, we must also accept that he had enough temporal judgment to know that he would be using that tool for some future hunt. Most likely, he was not just living in a perpetual *now* life. Preparation for the future is a talent that many living things have. Trees prepare for the seasons, the dryness, wetness, cold, warmth, and the height of the sun. Birds migrate, squirrels collect food, and most living things plan for breeding. Of course, in animals and plants, such preparation is mostly instinctual; in humans, it is more connected to consciousness. Without knowing for sure what went on in the mind of a cave-person, we can always imagine and speculate about his or her temporal acuity; however, we know quite a bit more about modern humans. We have the unique ability not only to think about the future but also to envision the future, and even to picture ourselves in that future. We have an undercurrent awareness of continuity and connections between temporal states that perpetually guard against threats to better conditions of life. We humans surround ourselves with goals for improvement, with a time-conscious awareness that infirmity, and good night, lurks at the end.

---

At some early point in history, human temporal awareness sharpened to accommodate the social and physical environment. As societies became more sophisticated, as clocks and calendars increasingly controlled the patterns of life, time tended to dominate cognitive actions of planning days, inescapably involving the temporal categories of past, present, and future.[8] Such planning sharpened the human concentration on present actions and the anticipation of future needs. The twentieth-century British priest and comparative religion scholar Samuel George Frederick Brandon, who once held that the fear of death as the ultimate end might have spurred the basic motivation for many of the world's religions, also said that "generally speaking, the most successful persons or societies are those that have been most capable in exploiting time-sense in the planning of their affairs."[9] He believed that awareness of time enabled the human species to dominate all others in the struggle for life.

Although two hundred thousand years seems like a long time, as a modern human species we may be a long way from our half-life. The job for humans is to stick around for a much longer time. A gauntlet of miraculous genetic accidents equipped the human brain with antennas that focus on what is coming and anticipate change with the power to guide and control ambushes of shifting environments and unexpected events. The human brain amasses information from surrounding environmental situations and uses it to anticipate possible dangers to the body. Danger is a notion that implies a detrimental something that has not yet happened, something that could or will happen in the future. With any kind of reasonable consciousness, the brain's anticipation of danger creates a future that is beyond the present. The brain distinguishes between past and future, cataloging events as they zip by into memory. True, we sometimes live for the day. Football players know the consequences of head injuries, and we sometimes eat things we shouldn't.

How do we get from simple instinctual responsiveness (omitting philosophers' ongoing debates about the realness of consciousness) to a sense of time, or at least to an awareness of the separation of past with present and present with future? The answer leans on uneven ground, because we know next to nothing about time awareness in precursors of conscious humans. We know just a bit more about time awareness in modern humans. The brain is very efficient in keeping time for its own functions. We know that raising an arm to throw a spear, for instance, starts with body-sense information arriving in the brain and leaving at an impressive speed to coordinate the commanded muscle functions. The batter who can spot a baseball coming at him at ninety miles per hour can have an amazingly coordinated response from his body and arms with a pinpoint accuracy of the swing of his bat. How does the brain coordinate such a tricky maneuver? The batter's swing didn't originate in the sport of baseball. Rather, it likely came from the hundreds of thousands of years when neurons grouped in new networks to enable the species to get food by hunting speedy wildlife. As the speed of gazelles increased, so did the neuromuscular coordination of humans, evolutionarily.

Some archaeological testimonies give us a conjectural and limited grasp of what triggered the first understanding of passing time as a division of present from past. Humans, like their primate ancestors, were, perhaps always, social beings. And although we can go back to a time when there might not have been a language beyond grunts of vowels, those early humans had a memory through pictures and, therefore, a distinction between past and present. The way we distinguish those temporal parts has something to do with language. Cognitive scientists Rafael Núñez at the University of California, San Diego, and his graduate student Kensy Cooperrider tell us through their research studies of the Yupno people in New Guinea that cultural linguistic metaphor plays a central role in how people conceptualize time. Núñez believes that one way of understanding time is through spatial metaphor, since humans are instinctively good at navigating and mapping space.[10] An English-speaking person is likely to refer to yesterday by suggestively pointing backward over a shoulder, and forward for tomorrow—the past is behind us, where we came from, the future is ahead, where we are heading. In American Sign Language a person would indicate the past tense by using the infinitive of the verb with a wave over a shoulder and the future tense with a wave ahead.

---

Time lives with me—always has.
Follows me, pushes me,
brings me to work—wherever.
Hangs by to wait for me,
lags behind flights west—always
catching up in reaching home
to know me intimately,
to show I'm not alone.
*—JM*

We should not mistake time perception for timing within the workings of the brain. Despite forty years of research, psychological, neurological, and pharmacological, we still have no definitive evidence of a connection between the brain's temporal mechanisms and the conscious perception of time's passing. The problem is that time comes disguised in many forms, one being how it behaves and another what it is. Many tests are done on animals under the uneasy understanding that humans do have different time mechanisms, producing conclusions that often do not reflect factual time in the real world. That apprehension comes from knowing that there is no direct connection between internal and external sensations that can be computed from any known instruments, PET scans, fMRIs, or even leading-edge neuroscience techniques of monitoring brain cells by introducing light-sensitive proteins. There is not one pathway carrying temporal information to and from the brain but many, enough to confuse the findings of neurological experiments.[11] Moreover, neuroscientists find that almost all sensory channels carry some instructions for time perception. And beyond that, the scales of time perception range from microseconds to years. The means by which the brain computes small or large durations are not clear.[12] And even more surprising, there are multiple mechanisms that "operate in parallel, complicating the search for simple information processing models and neural substrates."[13] There are some leads, though: dopamine can play a role in time perception, and unique parts of the brain are known to control circadian rhythms and dopamine delivery.

Years ago, I was one of several guinea pigs involved in a physics experiment to test perception in sensing short periods of light strobes. It was a terribly annoying ordeal, especially since it was voluntary. Two lights flashed in succession, and I was asked to count the number of flashes I had seen. I could count two. Gaps between flashes were less than a second long. The experiment repeated with shorter and shorter gaps between flashes. Always, the count was two. But at some point the number jumped to one. Even though there were actually two flashes, all I could see was one. When the gaps lasted less than about a tenth of a second, the flashes appeared together. It was as if the neural activity result of what I was seeing was experienced as a single moment of sight, as if the pair of events were packaged in one representative flash. Oddly, though, when the lights continued to flash at that short tenth of a second interval, I found that I could sometimes see two flashes, other times just one. It seemed that my neural time activity was a sequence of running packets and that I was packing pairs of flashes into packets of neural running time. On occasion, a pair of flashes would split so that each flash would find itself in two distinct time frames. Then I would see two, distinguishing one from the other. I was told that my experimenters believed that those time frames were neutrally ticking at rates dependent on dopamine levels: the higher the level, the lower the speed. This might explain why exciting events that produce dopamine alter our sense of time. The higher the level, the slower the internal clock relative to real time, and therefore the more one underestimates passing of time. This is to say that our sense of time is not constant and that dopamine activity bends time judgment.[14]

Time perception has been studied almost since the birth of psychology and cognitive neuroscience. The Russian naturalist Karl Ernst von Baer questioned what "now" is in his 1860 address to the Russian Entomological Society in Saint Petersburg. What is the moment between the past and future? He meant it to be the shortest moment conceived by humans. For Baer, that moment is different for different animals. But for humans it seems that the shortest duration that could possibly be part of the inner sense of a moment would be no shorter than forty milliseconds. It is what the German neuroscientist Ernst Pöppel at the Institute for Medical Psychology and Human Science Center in Munich University calls a time point, a "human moment," or a time quantum, a simple "temporal window for a primordial event which is a building block of conscious activity."[15]

Pöppel tells us that we experience time either as a duration or an overlap of durations (two or more events close in time) or as an ordering of events or as past versus present or as change. He curiously tells us that the events of past or present are different from the experience of order, the ordering sequence of which event came first. Try to capture the instant of now before it forever escapes reflection. "Let us try," the English philosopher Sandworth Hodgson proposed. "Let any one try, I will not say to arrest, but to notice or attend to, the present moment of time. One of the most baffling experiences occurs. Where is it, this present? It has melted in our grasp, fled ere we could touch it, gone in the instant of becoming."[16] For Hodgson, time is an abstraction that never really comes through the senses but seems to be always with us as part of what Robert Kelly, the builder of a cigar company who later in life became an amateur philosopher, called the "specious present."[17]

---

Hodgson was right. Yet I would go further. That specious present does seem to melt, but it does not totally disappear. Like dandelion seeds in a wind, experiences of sequential moments accumulate in memory to further human accomplishment. When you first learn to drive a car, you are keenly aware of everything that must be done, even though there are some things you are not aware of that might contribute to your driving skills. You are paying attention to everything being told. You are concentrating on how slowly or fast your foot must press on the gas pedal. You are thinking about how quickly that gas pedal foot will react in transferring to the brake pedal in case you might need it to transfer. You are intensely watching the road and how close you are to other cars and pedestrians. You do this for several days, perhaps weeks. Before long, the intensity of your conscious attention to detail moves to different locations in the brain to automate almost all your body's motor controls. Each time you drive, more and more of those controls become more and more routine while the motor limbs of your body—leg, hands, and fingers—seem to develop separate "minds" of their own. That initial conscious intensity is soon built into a nonconsciousness that permits you to relax and do other things that you could not have done on day one of learning to drive.

Learning to walk, to ride a bike, and to play piano needs intense concentration to enable success. You might think that walking is instinctual. But the child who is ready to walk, and who is teetering during his or her first few steps, is surely concentrating on how to balance and not fall, just like the child learning to ride a bicycle. The body seems to take over all those cerebral tasks that had to be coordinated in order to perform the task well. We, who already know how to walk, don't think about the steps we take to get from one place to another, though they may be complicated to the extent that the task might mean walking up or down stairs or running on uneven ground. Swimming requires learned skills that are not natural to land animals; yet, unless we are performance swimmers, we don't think much about where our arms and legs are when we swim for pleasure. Learning to play the piano is a bit different in that there are no frights that could happen (other than humiliation), but it is also similar in the intensity of concentration to succeed.

Musicians know something about finger memory. Play a new piece on the piano while consciously reading notes you've never seen before. Do this enough times, perhaps twenty, perhaps a hundred. Soon, your fingers are doing all the work. You might be looking at the sheet music, yet not fully concentrating on the notes and fingering. Your fingers might start to take over to let you concentrate on tempo, phrasing, and more creative interpretations of what you see on the pages. Part of the reason is that you hear the music in your head in advance of where you are. That music in the head is some milliseconds ahead of you, coming forward to meet your "now." You have advance awareness of notes, timing—yes, timing, especially timing—and phrasing heading toward you, coming out of the future, and somehow, all the routine motor repetition coming from endless practicing pays off in the form of motor-controlled music. That observation is probably very false for real musicians, but it is what the novice musician feels after spending endless hours practicing a simple musical score.

So it is with many of our early dealings with conscious activity. It is that way with almost all learning. And so it is with time. The sense of time, through the witnessing of it in so many disguises, in so many variations, and in so many circumstances, eventually gets distributed to various parts of the brain for more skillful knowing of time itself. It is as if separate parts of the body know things by themselves, yet accept prompts from other parts, and collectively compose a subconscious rhythm.

In old age some perceptions of time reverse. It's as if all those various parts of the brain that stored information to make us less conscious released all that knowledge to the winds. As we lose our sense of balance, we lose our sense of time as well. The irony is that as time becomes more important, our sense of time becomes looser. We have fewer things to do, fewer obligations, and more time to fill. But we also become slower at doing things and less anxious about not doing things on time. It seems that time should become more important, but in reality it becomes less so as we approach those ages when the days of life become more precious.

## INTERLUDE: UNDERCOVER AT AN IPHONE ASSEMBLY PLANT IN CHINA

In the United States there are few jobs as boring as that of an assembly-line employee screwing caps on toothpaste tubes. Now such jobs are only imagined through the fictional Mr. Bucket's job in Roald Dahl's *Charlie and the Chocolate Factory.* Machines and robotics have largely taken over those inhuman jobs of the past, leaving other occupations that are dull and tedious.

In China, Vietnam, and Cambodia, things are different. Hundreds of thousands of people are still doing assembly-line work more monotonous than Mr. Bucket's. Speak to Dejian Zeng, who spent six weeks working in a factory near Shanghai assembling iPhones. Zeng was not putting lots of parts together, as if he were building Lego models of the *Star Wars Imperial Star Destroyer;* rather, his job, all day long, was to insert and tighten one tiny screw over each iPhone speaker that came along the assembly line so as to fasten the speaker to the housing. That's all he did for twelve hours a day, six days a week. The only breaks were ten minutes every two hours, a fifteen-minute lunch break, and a thirty-minute dinner break. Zeng was doing investigative work for a group of researchers at New York University, where he was a graduate student specializing in human rights in China.

In the beginning, Zeng had a hard time catching up with the speed of the assembly line. It moved fast, so he had to remain focused on the simple task, which was extremely tiring. After a while the robotic technique set in, so he could perform the mindless task of the simple job with his eyes closed. At that point, his mind felt blank. Electronic devices of any sort were not permitted. He told Kif Leswing, a reporter for *Business Insider*, that the mindless work slowed time almost to a halt. "What I did," he told Leswing, "is that I put the speaker on the case, and I put a screw on it. The housing—we call it the back case—is moving on the assembly line, and that's when we pick it up, and now we get one screw from the screw feeder, and then we put it on the iPhone and then put it back, and it goes to next station."[1]

There were more than a hundred stations, each responsible for just one specific part of the assembly. It is hard for most people to imagine the tortuous thinking about time while doing one mindless task like securing a little screw in an iPhone. This kind of labor is so foreign to what the human body has experienced. There is almost no other labor to compare it to for the best example of extreme boredom. Machines now pick cotton, gather fruit, make blue jeans, assemble cars, and even place caps on toothpaste tubes. However, the iPhone needs a tiny speaker screw to be hand installed. Someone's gotta do it!

Workers were paid about four dollars per day. One would not be surprised if the Pegatron factory might consider a robotic food machine to feed workers lunch without stopping production. Time is continuous, so why not production?

Zeng's story is reminiscent of Charlie Chaplin's Tramp, who works on an assembly line for the Electro Steel Corporation in the film *Modern Times*. The Tramp uses both hands to tighten a pair of bolts on each fast-moving part coming along the line. Along comes the Billows Feeding Machine Company to demonstrate its newest machine, which will help produce more widgets by eliminating lunchtimes. *Modern Times* is a satirical silent movie, so we don't know what the Electro Steel president is thinking, but surely he would love to eliminate half-hour lunchtimes, boost production, and keep workers working as they eat. Why not? It translates to three weeks of free labor per year. For a large company, that's a lot of money saved for the owners. Unlike money, which could accumulate exponentially, time accumulates linearly. My summer job at my uncle's silk-screen shop that lasted just one week showed me just how inhumanly boring a job can be. Assembly-line jobs in America these days are rare. Almost every manufacturing job is now automated, with reliance on some human watching for tainted or damaged products that inevitably escape accepted standards of assembly.

PART

# LIVING RHYTHMS

Imagine the body as a Rube Goldberg machine, with thousands of tiny devices whose cogs, baskets, and springs must align correctly in a moment for life to proceed. And it turns out that not all the springs or baskets are present at any given moment. If you send a marble down a chute, the route it takes in the morning may be different from the route it takes in the evening.
—*Veronique Greenwood, "The Clocks within Our Walls"*

# 15

## THE MASTER PACEMAKER (THE EYES OF TIME)

> Lately the subject of my thought is time:
> How time flows like a river, if it does,
> Or like a house of crystal, stands immobile,
> Or ceases where the very smallest creatures
> Dance without houses, clocks, or bankless streams.
>
> Here is my birthday, carried round again
> By the tall star-crossed cycles of the moon.
> Love's shadow brightens as the day begins
> And then grows longer, whether we believe
> That time is real or just a dream of things.
> —*Emily Grosholz, "Love's Shadow"*

The focal home of the body's internal clock is in the hypothalamus, almost exactly, but not quite, where René Descartes thought the seat of consciousness would be, next to the pineal gland, which he considered the principal seat of the soul. For Descartes's mind-body dualism, the input of information comes through the eyes and is passed on to the pineal gland in the center of the brain to contribute new information to the consciousness of self and mindfulness.

We now know that consciousness is distributed through many regions of the brain. We also know that there cannot be just one single moment when a conscious awareness of any decision will occur. The alert brain is constantly entangling myriads of experiences bombarding the central nervous system and therefore has no one precise area that can be tagged to the consciousness of a particular event.

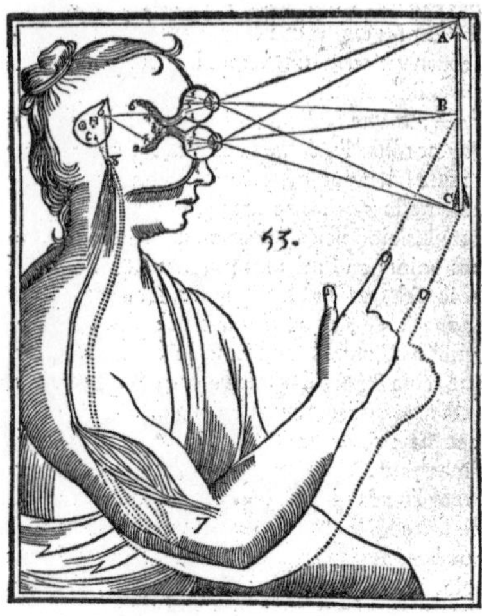

René Descartes's illustration of coordination of muscle and visual mechanisms, as drawn by Claude Clerselier, editor of Descartes's *Traité de l'homme* (Paris, 1664). Wellcome Collection. 

Descartes created a comprehensive analysis of the human eye by conducting experiments on an ox. He was twenty-four years old when he learned eye physiology from Christoph Scheiner's *Oculus hoc est: Fundamentum opticum*. Scheiner was a Bavarian Jesuit priest physicist who conducted experiments on oxen eyes to learn more about the anatomy of the human eye. Descartes extended Scheiner's experiments by removing the back of an ox's eye and replacing it with an eggshell. He made a small hole in the back of the eye, taking care not to damage the intraocular fluid of the eye, and then placed the eggshell over the hole. Images placed in front of the eye were reversed on the shell. He learned that images are inverted on the retina in the back of the eye and that the brain receives images through ocular nerves.

In this way Descartes was able to describe the function of the iris and the ring of smooth muscle in the eye's vascular layer, a muscle that has many functions as well as control over the shape of the lens for viewing objects at different distances. Descartes was interested not only in the physiology of the eye and its components; he wanted to know how light in the eye and hence how images that were cast through light-sensitive cells (which, of course, he didn't know about) in the retina got transmitted to become what we call "seeing" in the brain. For a seventeenth-century experiment that question was old, but far too complex to answer without knowing something about the existence and function of optic nerves and without knowing that chemicals and electricity trigger nerve impulses relayed to several processing centers in the brain. The retina contains layers of light-sensitive tissue. Whatever light is detected is transformed into electricity in much the same way that light falling on a piece of silicon (a solar collector) creates an electrical charge. Light-sensitive tissue sends a signal to the brain to inhibit the release of melatonin, the hormone that chemically lowers body temperature and causes drowsiness. In darkness, melatonin is normally released.

In the Cartesian view of dualism, we have eyes, ears, and brains on the side of the physical. Eyes and ears take in information, as Descartes's illustration suggests, then pass that information to the pineal gland in the center of the brain, headquarters of imagination and common sense, as well as a door to consciousness. It looks like a small pinecone and is located just between the two halves of the thalamus. The human pineal gland might have had some direct photosensitivity at one time in its evolution timeline, now gone and replaced by synaptic information coming from photopigment in the retina.[1]

Is there really a dedicated internal clock that is coordinated to a centralized system for measuring time coming from different stimuli and different tasks, an endogenous pacemaker that can tick away, detect, and control the pulses of the body's rhythmic system, and synchronize internal biological rhythms with external environmental cues of daylight and seasonal cycles? Yes.

We organize our days mostly according to habits and duties. There is a meeting, a deadline, work to be done, a clock to be watched. The body's needs are synchronized as a clock with the alternating rhythms of the surrounding world. The focal home of that clock happens to be in the hypothalamus. Whatever that clock actually is, its function is to control and regulate body time.

Aside from the human temporal processing system concentrated in the cerebral cortex and basal ganglia, there is also the suprachiasmatic nucleus (SCN), a pair of nuclei in the hypothalamus at the center of the brain, a place very close to the pineal gland, which produces melatonin to regulate sleep patterns and body temperature. That SCN is believed to be the master pacemaker somewhat responsible for the control of circadian rhythms. Many organisms, from the simplest single-celled algae to humans, have twenty-four-hour rhythms based on genetically based feedback loops. Those loops tell the organism when it is time to sleep, time to wake, time to change body temperature, and when to be alert to dangers. The SCN can sense all sorts of body functions that could interfere with the synchronization of internal and external time by receiving information from the eyes indicating that light is in the environment. For instance, hunger, nourishment, anxiety, health, and work schedules can offset the clock. So the SCN is the pacemaker that synchronizes internal biological rhythms with external environmental cues of daylight and seasonal cycles. It can help change the concentrations of melatonin so that the molecular workings of the internal clock are coordinated with the earth's twenty-four-hour rotation. There is ample evidence for this. When a healthy mammal SCN with functioning circadian rhythm is transplanted to another animal suffering from arrhythmic behavior, rhythmic behavior is restored.[2] Moreover, SCN neurons stored in cultures continue to follow circadian rhythms.[3]

So Descartes was on to something. There *is* an internal clock, and indirectly, it has something to do with photosensitivity through the retina. But Descartes was also wrong about something. There is no particular nerve center that sparks consciousness. For him (and for many of us, too, for that matter) the mind is different from the brain—different in the sense that the mind is composed of something different from what we normally call cell tissue. That is the theory of dualism—there is both a material brain that takes in information but does not think and a mind that brings consciousness to the choices a person must make to benefit the self. Materialism counters dualism in that it believes that the mind is purely physical, made only from matter, with no help from some sublime transcendental substance (possibly coming from one's immortal soul) that does not obey the general laws of physics, chemistry, and physiology. In other words, we should be able to account for all mental processes according to the physiology of the material substance of cell makeup that can either be seen with the help of detecting instruments or surmised through the laws of physics, biology, chemistry, or some scientific discernment. No sublime nonmaterial substance is allowed. We are pretty comfortable with the notion that the body is a collection of physical materials, cells of all kinds that divide, grow, and die. But when it comes to the mind, our beliefs tend to scatter to attitudes suggestive of duality. Some people do not accept the thought of materialism because there is so much that it cannot explain. It promotes what has been called the mind-body problem, the problem that begs the question of how mind and brain communicate to provide mindfulness.

---

The word *mindfulness* comes wittingly from the English translation of the Buddhist term *sati* or the Sanskrit word *smriti,* whose root is *smara,* which means everything between remembering and thinking. The closest common word in English is *awareness.* But mindfulness in the Buddhist philosophy is more than just awareness and remembering of sacred texts, events, and environment. It is one of the seven factors of enlightenment that now involves awareness of one thing in relation to another and, more than that, an awareness of awareness.

Mindful time is the mechanism by which we experience the passage of time and how we consciously think about it. It is not a clock. A clock is usually consistent in its metronomic ability to keep the beats of *tickings* and *tockings* in regimental uniformity. The mind clock is, for sure, something organic that has—as do all organic substances—a built-in irregularity, a living metronome that slows down and speeds up in favor of the body's best chances of survival. It is likely that the mind itself—if, for the moment, we dare to separate mind from body—has no direct connection to an organic timepiece but rather has some indirect connection with the conscious attention to the town clock—that is, to the collective precision of time in our neighborhood and environment. That environment shapes our circadian rhythms, those biological processes that repeat and maintain roughly twenty-four-hour rhythms in the absence of any external signals. The SCN uses more than twenty thousand neurons to synchronize body rhythms by a combination of internal and external cues, with the eyes performing a small role in controlling and regulating body time.[4]

Human eyelids play a central role. Yes, it comes down to the natural selection of eyelid function. They are 20 percent transparent. Morning light is detected by ocular nerves to signal the body that it is time to wake. This is a convenient hangover from primitive times that relied on light sensitivity along with cocks crowing to separate the sleeping body's dis-

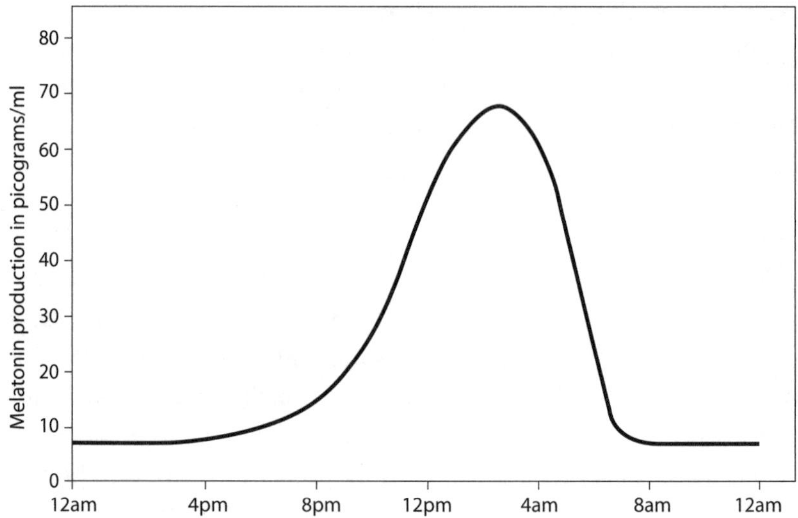


Melatonin production cycle during summer months

tinction of day from night. That eyelid transparency helps the internal clock, whatever that is, to calibrate with the normal functions connected to day and night distinctions. An opaque lid would have us sleep without wakeup signals to the brain. Inner clocks of people living in the far north or far south calibrate day from night by other means. Of course, with total blindness, other factors kick in to synchronize biological rhythms, and melatonin is produced in rhythmic cycles by other means.

Changes in light and darkness sensed by ganglion cells in the retina send that information along the optic nerve to the SCN. Ganglion cells in the retinas contain a short-wavelength light-sensitive pigment that even many blind people are sensitive to. Some signals arrive at the pineal gland. When retinas are exposed to light, those signals are suppressed; in darkness they are activated. In this way the circadian clock is reset each day. From the graph shown here, we see that in summer months, melatonin production begins at roughly 8:00 p.m., peaks at about 3:00 a.m. to give the deepest sleep of the night, and ceases at about 8:00 a.m.

Why do we spend so many hours of the night sleeping? On the one hand, almost every organism on earth is in lockstep with the harmony of the planet's day and night dance. On the other hand, many animals do not sleep at all. Dolphins shut down one side of their brains and one eye, leaving the other side and other eye to watch for predators. They swim while literally half asleep. Whales and porpoises do the same. Giraffes take five-minute naps about six times in the twenty-four-hour day. They nap while standing, at times resting their heads on their backs. They stand ready for that lion pride on the hillside. That's the extent of their sleep needs. Evolution has given them the advantage of being tall to reach over the tops of trees, along with the disadvantage of not being able to rise quickly from a lie-down position when lions are nearby. One would expect that elephants would need lots of sleep, given their gargantuan size. But they sleep for no more than thirty minutes at a time, most often standing. The Alpine swift migrates from the Himalayas to South Africa, taking six months in an almost continuous flight without a break for sleep. Some research suggests that it might be asleep along some parts of its flight, with outstretched wings guided simply by the currents.

These animals seem to have escaped the need for long periods of sleep. Their sleeping habits do not conform to the dictates of light and darkness. Perhaps sleep is not a matter of fatigue. The question here is whether the external twenty-four-hour clock has strong, weak, or no influence on the human internal clock that dictates so many of our body behaviors, such as temperature variations (the phenomenon of internal temperature being high in the evening and low in early morning). Internal temperature, tuned to a twenty-four-hour rhythm, is a physiological condition that is also found in people living in latitudes where daylight and darkness shift from one extreme to the other over summers and winters, and wherever sleeping habits differ from the typical rhythms of people living at lower latitudes. Even with the large amount of data collected from laboratory experiments on biological rhythms, the conclusions are not precisely clear on the question of the external clock's influence on the internal clock.[5] However, current data suggest that physiological rhythms (the circadian rhythms) in humans are synchronized by light, even for those who became blind long after birth. It seems as if the habits of synchronization continue after photoreception sensitivity ends.[6]

# INTERLUDE: TIME ON THE TRADING FLOOR

A New York hedge fund trader once told me that time is money. "Hedge funds are just cheap bets on credit," he admitted. "Every day, I put money—millions of dollars—on futures and credit-default swaps. It's a legal crapshoot, but you gotta know what you're doing. I can lose a bundle in one second, and make lots of money in the next."

I thought about what that trader meant by that last sentence, and then imagined the amounts Bill Gates, Jeff Bezos, Mark Zuckerberg, George Soros, Michael Bloomberg, and Warren Buffett gained or lost in just fifteen seconds.

The digital clock on the trading floor of the New York Stock Exchange is somewhat precise, yet not connected to the trading bell. The opening bell of the exchange is rung by a human at almost precisely 9:29:45 a.m. Eastern Standard Time by the push of a button and again for closing at 3:59:45 p.m. EST. Those fifteen seconds could significantly shift some dealing in a typical day's five million trading transactions worth forty billion dollars. High-speed computers trade thousands of shares per second, and even relatively slow floor trading can result in a shift of tens of thousands of dollars in just fifteen seconds. Such high-frequency trades placed from locations all around the world account for almost 70 percent of the trade volume in the United States. According to *Forbes,* the net worth of Jeff Bezos and his family in 2019 was about $131 billion, of Bill Gates about $96.5 billion, Warren Buffett about $82.5 billion, Mark Zuckerberg about $62.3 billion, and Michael Bloomberg about $55.5 billion.[1]

These are staggering amounts of money. Just to give you an impression of how large these numbers are, consider how much money is gained or lost in just fifteen seconds of a normal day when the stock market is open. If we assume that as little as 30 percent of Bezos's money is tied up in securities and exchange investments, Bezos could either gain or lose about $4.4 million every fifteen seconds, on average. So time really is money, and for some people it is really big money. For Bezos, $4.4 million is not much.

# 16

## INTERNAL BEAT (CLOCKS IN LIVING CELLS)

> The funny thing about life is that it's temporary; that is to say, temporary in the "temporal" sense of the word, meaning that all living things and all that we do are subject to the precepts and effects of time.
> —*Lansing McLoskey,* Theft

Many organisms perform best at certain hours of the day. The slug species *Arion subfuscus,* living in almost total darkness, knowing nothing about the Gregorian calendar, lays its eggs between the last week of August and the first week of September.[1] Bees forage for nectar, knowing the best times to visit the best fields and the exact timing of nectar secretions for individual species of flowers.

In the mid-twentieth century, the Austrian Nobel laureate Karl von Frisch provided enormous insights on honeybee communication and foraging time. He discovered that bees have internal clocks that tell them not only where the nectar is to be found but also precisely when that food will be ready. "I know of no other living creature," he wrote in his book on bee language, "that learns so easily as the bee when, according to its 'internal clock,' to come to the table."[2]

Indeed, honeybees start their daily routines of harvesting nectar by the clock or, rather, by sun time. While studying bee routines at his lab at the University of Munich before World War II, Frisch trained bees to come regularly to lunch at strategic times when feeding stations were set up with sugar water. The bees quickly adjusted their natural schedules to

Frisch's artificial schedule. In just two days their old schedule was abandoned. Even informational nectar-finding flights were stopped.

Judging from the conclusions of Frisch's experiments, it seems as if physiotemporal processes in animals and insects rely on some internal rhythm that we could call an internal clock. Whatever it is, it must have some connection to the external events of sunlight and moonlight, the local time of earth's turning and orbiting. Frisch's student Martin Lindauer later corroborated and advanced Frisch's experiments by hatching bees in a controlled environment of twelve-hour daylight and twelve-hour darkness. When the bees reached maturity they were trained to head in a given direction for at least five days.[3] They learned the time of day purely from the sun's position.[4] Lindauer and Frisch were very much surprised, since at that time it was well known that birds use innate migration routes to navigate their journeys, and so do many insects. But a bee's journey is more complex because it changes from day to day.

---

"There is another tree with many leaves like the rose, and that closes at night, but opens at sunrise, and by noon is completely unfolded; and at evening again it closes by degrees and remains shut at night, and the natives say that it goes to sleep."[5] This quotation is a translation from the Greek that was written by Theophrastus of Eresos in the third century BC. He was describing the daily leaf movements of the tamarind tree, suggesting that it is an organism whose physiology follows and responds to the time of day without any external cues, such as exposure to sunlight.[6] Of course, we know now that all organisms have zeitgebers to help them adjust to seasonal variations of daylight, and so the tamarind surely had help from hidden cues. Buds appear without fail in the first week of April on a lilac bush on the south side of my house. The temperature could be close to freezing, and yet those buds will poke through as if to say, *I sense the sun's altitude angle is at about 51 degrees at noon, and that's enough daylight for me to come out. Yay! Spring is coming! Now, where are those bees?*

Experiments performed by the French astronomer Jean-Jacques de Mairan in 1729 established that plants have amazingly precise rhythmic behaviors that are independent of their environment. Mairan was interested in leaf movement and why certain plants, such as *Mimosa pudica,* a perennial weed found mostly in Central America, South America, and Asia, spread their leaves in daylight and fold them at night. Mairan recorded the mimosa's leaf behavior under controlled conditions of constant darkness. Every day, in total darkness, the *Mimosa pudica* leaves opened during daylight hours and closed at almost the same exact time every evening. Much is now known about these orderly rhythms. First there is the sensitivity to temperature cycles. Mairan's experiments with light did not account for temperature and moisture variations. The French physiologist Henri-Louis Duhamel du Monceau, who repeated Mairan's experiments in 1758 with as much control as possible, placed the plants in blankets in a very temperate wine cellar. Despite all these isolations, the plant movements continued.[7] Even without a light clue, the plants were able to tell time.[8] In 1832 the Swiss botanist Augustin Pyramus de Candolle discovered that after the *Mimosa pudica* remained in darkness for a few days, its leaves would open about one or two hours earlier. The plant seemed to be adjusting its rhythm to its old routines with the sun, but its leaves' sleeping period never went below a twenty-two-hour cycle.[9] Later in that century, Charles Darwin also got into the act while writing *The Power of Movement in Plants,* a book about leaf exposure to the sun and night. He wrote that natural selection favored the ability of these "sleeping plants" to protect themselves from the cold of night with the benefit of sun absorption during the day.[10]

Even with the support of Monceau and Darwin, biologists did not fully embrace a belief in endogenous clocks. Some asked for a cause. Others asked for more convincing evidence that unforeseen environmental factors were not contributing to the phenomena. Then, in 1930, the German biologist Erwin Bünning experimented with *Phaseolus,* a native American wild bean plant, in total darkness and in total control of uniform temperature. He found that *Phaseolus* had a twenty-four-hour cycle that did not coincide with any variations of the environment; therefore, the plant had an endogenous clock mechanism that controlled the folding and unfolding of its leaves. He experimented with red light to study its effects on leaf movement in marigolds and found an extremely accurate daily rhythm without any apparent cue from their environments. It seems that some plant rhythms have an inherited internal time scale that either obeys or competes with day-to-day light durations.

---

Insects, too, have internal daily rhythms. If your home has ever been invaded by fruit flies, you know how annoying those tiny buggers can be. For every one you whack or crush, ten more will find you. What could be interesting about them? How much information about the world could come from those minuscule organisms the size of a poppy seed? It turns out that fruit flies share a significant percentage of genetics with humans, so they are model organisms for studying human-disease genes.

The complete mechanism of fruit fly genetics is as dauntingly difficult to explain as the depths of string theory. Fortunately, the outlying structure of at least the insect's circadian clock can be fairly well understood without scaling the endless scaffolds of biochemistry, endocrinology, and physiology supporting the common genetics of fruit flies and humans. Early fruit fly molecular research at Cal Tech is documented in Ronald Konopka's and Seymour Benzer's landmark 1971 publication, which reports success in generating mutant genes that became arrhythmic.[11] Leading chronologists claim that the work of Konopka and Benzer had an influence on the field of chronobiology that cannot be overstated, and their conclusions "were prescient for the entire circadian field and all of its subsequent molecular sophistication."[12]

The simpler story, told here, is sure to be considered raw to the biochemists and entomologists that call the fruit fly a *Drosophila melanogaster*, though I expect it is the best that can be done within these few pages. So, without getting into the full description of the differences between protein functions of humans and flies, we can interpret the *Drosophila* model of the circadian clock as simply a feedback loop that operates by a specific gene expression with a relatively short half-life. In essence and in generality, the loop simply behaves like this: the quantity of $A$ molecules increases, reaching a threshold that creates $B$ molecules (with a relatively short half-life), which in turn shut down production of the $A$ molecules.

Unlike fruit flies, humans have strong temperaments and wills that

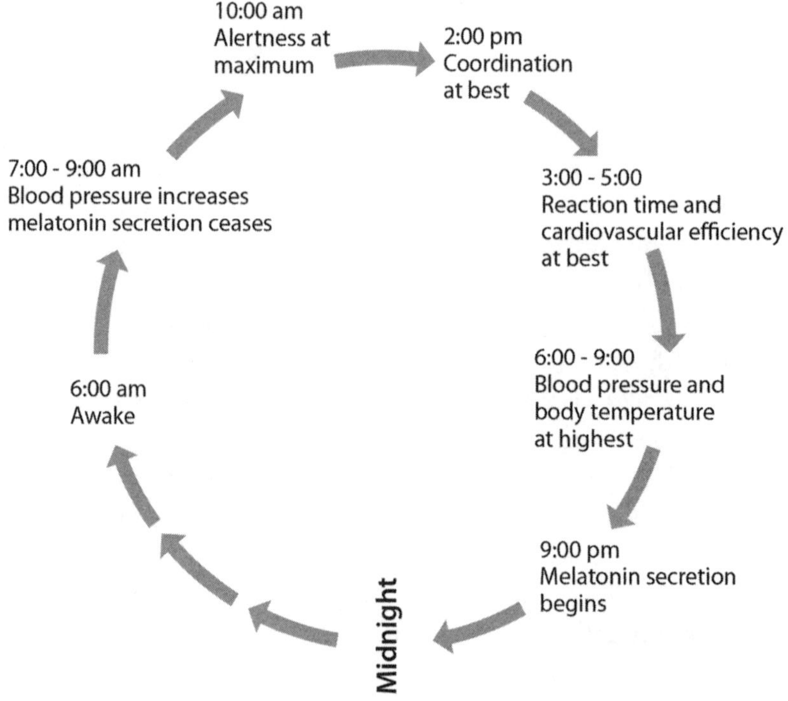


Circadian rhythm of the human biological clock for someone who wakes at 6:00 a.m.

permit defiance of their weaker, yet persistent, biochemical controls. The model circadian rhythm for a person who rises at roughly 6:00 a.m. and runs through each day with a regularity tuned to the sun is broadly illustrated above.

The human mind and body have a built-in circadian system, a coordinated assembly motivated to perform certain tasks, what we might call the macrobiological clock. At the molecular level there is a circadian oscillator, specific cell groups that work together, like the mechanism of a clock, to cause a larger system of mind and body to function in a daily rhythm. In the early 1980s Jeffrey Hall and Michael Rosbash, along with Rosbash's graduate student Paul Hardin at Brandeis, discovered such a circadian oscillator in the fruit fly, an insect that has timekeeping gene qualities associated with clock genes in humans. Hall and Rosbash won the 2017 Nobel Prize in Physiology or Medicine for their discoveries of the molecular processes that control circadian rhythms. In their seminal paper in the *Proceedings of the National Academy of Science,* they isolated the so-called *Period* gene (*per*), which cycles the amount of messenger RNA (mRNA) produced in a feedback loop, first forming and then terminating proteins made from *per* gene instructions.[13]

For clarity, let's first briefly recall the mechanisms of mRNA and proteins. The primary function of a gene, which is after all just a bit more than housing for a segment of DNA, is to give instructions for making protein molecules. Proteins, those chains of amino acids (chain groups containing oxygen, carbon, hydrogen, and nitrogen) responsible for maintaining and repairing cells, are made from genetic instructions transmitted from DNA to the ribosomes by messenger RNA. Ribosomes are intricate molecular machines that link amino acids together in the order specified by those instructions. The DNA in a cell's nucleus stores all the gene instructions for copying specific segments of DNA into RNA that are essential for continued life. The mRNA within the nucleus leaves the nucleus and enters the cytoplasm to dictate information stored in the genes.

All living things learn to manage daily environmental changes, especially the atmospheric lightness and darkness caused by the twenty-four-hour cycle of earth rotation. A human's hereditary information includes the biochemical mechanisms of proteins gotten from the routines of his or her ancestral life. And although millions of cells in a person's body have specialized functions, every one of them contains the same code of hereditary information.

Ever since 1992, when Hardin, Hall, and Rosbash published their findings on circadian oscillations, it has been known that fruit flies can tell time by their *per* gene's instructions. This led to the idea that a circadian gene's instructions are responsible for *per* mRNA cycling, and, by a feedback loop, *PER* protein also acts with return response instructions. The magic here is that the *per* gene located on the X chromosome of the fly cell contains the information for the mRNA (that has a relatively short half-life), which instructs the ribosomes to produce proteins connected with the *per* gene (called *PER* molecules, capitalized to avoid confusion with the *per* gene) that effectively travel back to the cell nucleus to turn off the activity of the *per* gene. Morning light would then destroy *PER* molecules. With *PER* molecules gone, the *per* gene would renew the process of encoding mRNA

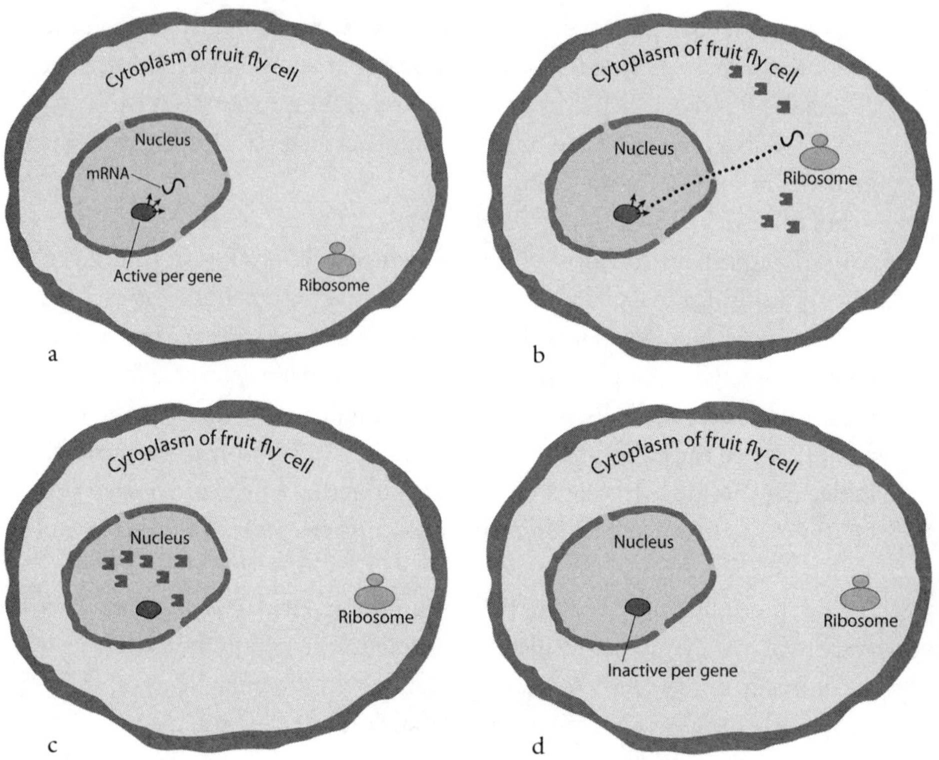


Molecular instructions of clock genes and proteins during *Drosophila* circadian rhythms: (a) late evening, (b) nighttime, (c) morning, (d) daytime

to complete a twenty-four-hour feedback loop. In effect, it is the fruit fly's molecular clock hand encapsulated in a single cell; moreover, it has since been discovered that the biological clock in most mammals works by the same feedback loop, though in mammals it takes a whole group of *per* genes for the process to continue. It could be that this *Drosophila melanogaster per* gene model is the result of organic evolutionary adaptation of the earth's circadian environment to maximize survival and well-being on a planet where life existence is governed by the alternation of light and darkness.

Here is how the circadian oscillator of *Drosophila melanogaster* works. The *per* gene in the nucleus of the cell transcribes mRNA molecules that migrate to the cytoplasm to give information and a green light for ribosomes (the protein workshop) to build both stable and unstable protein

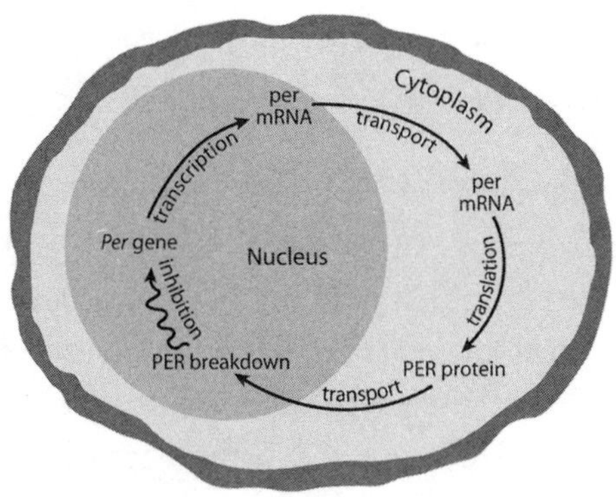


Twenty-four-hour feedback process of PER protein oscillations in fruit fly cell

molecules. The stable protein accumulates in the cytoplasm. As night goes on, protein levels accumulate to reach a threshold by roughly the middle of the night, when they enter the nucleus and begin to repress transcription from *per* gene instructions and soon after completely turn protein building off. In the morning, as the sun rises, the proteins decay and after several hours vanish. With all protein gone from the nucleus, the *per* gene turns on to restart transcription, and so the approximate twenty-four-hour-cycle loop begins again. On and on it goes, indefinitely.

The frequency of oscillation is entirely controlled by the rate at which proteins accumulate in the cytoplasm, the rate at which the entire threshold group of proteins move to the nucleus, and the rate at which proteins break down once in the nucleus. Those rates match up to provide a full twenty-four-hour loop, even without external cues. Perhaps that is why fruit flies hatch in highest numbers at dawn.[14] And also perhaps, this is why human sleep patterns tend to conform to ordered sleep times and why any disruption of the order creates confusion in subsequent sleep periods.[15]

Many living organisms, from snow fleas to *Mimosa pudica*, have evolved internal clock mechanisms that synchronize behavior, metabolism, and physiology with circadian rhythms. Humans, too, have specific cells that work as clock mechanisms tuned to circadian rhythms, but their oscillators are far more complex.[16] Here is what is known. First, the *Drosophila* model is now acknowledged to parallel molecular mechanisms in humans, and since *Drosophila* and humans have functional homologues of most parts of human disease genes, the *Drosophila* model is enormously effective for research into human diseases and drug discovery. It tells how the cell is attached to the night-day cycle of sleep, melatonin and endocrine activity, cardiovascular changes, body temperatures, blood pressures, immune differences, and renal functions.

Humans are programmed to be diurnal, active in the daytime and inactive (or less active) at night. That's a twenty-four-hour behavioral and physiological rhythm that expects environmental conditions associated with the rotation of the earth. Of course there are larks and owls because we are, after all, humans, all a bit different and off the charts of rigid controls. In 1994 Joseph Takahashi and his lab teams at Northwestern University and the University of Wisconsin used the fruit fly model to search for time genes in mammals, finding and identifying them in mice.[17] They identified the gene and named it *CLOCK*.[18]

Ensuing work established roughly homologous parallels between *clock* cell mechanisms in fruit flies and mammals. Mammals have three homologues to *per*, two of which produce *clock* proteins. Here is where things get too complicated for the scope of this book.[19] However, there are some reasonably easy things to be said about mammals that should help us appreciate the bigger picture.

In 1972 the neurologists Robert Moore and Nicholas Lenn at the University of Chicago used amino acid tracers to identify a light-information route from the retina to the SCN, suggesting that light cues entertaining circadian rhythms came from that pathway.[20] In that same year Friedrich Stephan and Irving Zucker at Berkeley showed that lesions in the SCN of rats disturbed their circadian activity and drinking rhythms.[21] Seven years later Shin-Ichi T. Inouye and Hiroshi Kawamura's work with rats at the Mitsubishi-Kasei Institute of Life Sciences in Tokyo confirmed without doubt that for rodents, electrical activity in the SCN followed circadian rhythms and that the SCN nucleus is an autonomous circadian pacemaker playing a major circadian role.[22] Another confirmation happened in 1990, when Michael Menaker and his lab at the University of Virginia transplanted an SCN from a mutant golden hamster with a considerably shortened circadian rhythm of twenty hours and twelve minutes to a nonmutant hamster missing its SCN. When the receiving hamster's circadian rhythm registered at twenty hours and twelve minutes, it became demonstrably clear that the SCN contains the pacemaker of mammalian circadian rhythms. The new rhythm could only be attributed to the transplanted SCN. In fact, restored rhythms seemed always to maintain the donor frequencies.[23] The role of ocular photoreceptors and synchronized endogenous rhythms stemming from light signals transmitted to the SCN became indisputable.

We now know that all cells, including those deep in the center of the body, can maintain an autonomous frequency of oscillations. Each cell holds its own clock, and each of those clocks is driven by the same feedback loop that guides the pacemaker clock in the SCN. In nonvertebrate mammals that pacemaker indirectly detects neural signals of light and darkness from the eyes to play a central role in regulating the body's circadian activity rhythms through sleep-wake cycles.[24] So the full body-clock system involves both the SCN in the brain and the trillions of peripheral clocks embedded in almost every cell of the body. We are a bundle of clocks that are synchronized to the environment by zeitgebers (about which more in chapter 18), of which changes in light and darkness are but just one.

With exception to cells in the eye, mammalian cells have no photoreceptors, so only the SCN can indirectly sense light by way of neural tract signals coming from the retina, and therefore, we tend to be awake mostly when light signals tell us to be. With normal and regular sleep-wake schedules the feeding cycle helps to align hormonal activities and to synchronize clocks in liver and intestine cells. Under noncircadian or restricted feeding schedules, the clocks of liver cells simply follow the feeding cycle, ignoring SCN calibration signals.

We might think that we have control over time's grip on our will and behavior, that time has some feeble influence over body functions vulnerable to willful suppression, and that biological links from the mind and body to the external circadian cycle are too delicate to be taken seriously. Not true. The biochemical and genetic structures seem to be stronger than we suppose. They dictate time, by way of cell-to-cell transcription/translation feedback loops under endogenous control, in synchrony with extracellular geophysical cycles, to bring about changes in the behavior of the entire organism. Surprisingly, the loop continues through a feedback circle from behavior back to the molecular clock.[25]

The healthy human body has many feedback loop mechanisms that signal functional information, from when to stop eating to when to rest. If we eat too much, the hormone leptin (an energy regulator) is produced to trigger a bloated feeling. Cells need to absorb nutrients, and they do so in a circadian rhythm of taking in nourishments by day and shutting down at night. In the course of a life, cells die and are replaced. That sequence happens with scars of the skin. Small rips of tissue are replaced with regrown cells. The same happens with cells well within the interior of the body and organs, where the breakdown and replacement cycle understandably coincides with the rhythm of sleep and wake times.

That breakdown and replacement cycle is just now being understood in the drug industry, where medications are being discovered to have optimum effects at particular periods of the twenty-four-hour day and to have catastrophic effects at other periods. Like trains, our bodies run on a timetable. The consequences of the rhythmic nature of individual cell functions could be a direct benefit for medication management and cancer radiotherapy.[26] Particular timings could enhance or diminish or rescind the efficacy of a medication and in some cases could result in serious danger to the patient, possibly even death.[27] Such seriousness has spawned a fresh pharmacological field, *chronopharmacology.*

This all stands to reason, since those body enzymes in the liver that break down drugs are abundantly produced according to habitual eating schedules.

## INTERLUDE: LONG-HAUL TRUCKERS

At a truck stop on I-91 in Massachusetts I met Rodrigo Rigarro, a long-haul trucker who works for the transport company J. B. Hunt. Rigarro had just come from Atlanta in an eighty-foot Peterbilt 539 heading to Montreal. He is a heavy-set fellow with a wave of salt-and-pepper hair tied in a ponytail. I sat next to him at the lunch counter, hoping to ask him about how time passes when he's on the road.

"I really don't think about it much," he said, before saying a great deal about it. "It really depends on the road. There are roads that are *hell-boring*. You can't wait to get out of the monotony of straight, flat road with nothing but corn on both sides. A bridge to cross would help."

Rigarro said he drives more than three hundred thousand miles a year. For him, every road is boring, but most of all the I-40 out of Knoxville. He listens to radio—pop and country music and news—and spends his time thinking of retirement, his next meal, and a shower. After a few minutes picking up the hardly touched part of his cheeseburger, Rigarro became interested in what I was writing. I was taking notes while also nibbling a rather dry hamburger. At first, he didn't want his name used in anything printed. He was afraid of saying something that J. B. Hunt would be unhappy about. But when he learned that I was interested in time and not in boredom, he articulated a flood of thoughts.

"I've thought about time," he said, contradicting what he said earlier. "I look at the clock, I look at my watch, I look at my phone many times while trucking. My GPS tells me how long it will take to get from Nashville to D.C., and I'm amazed when I get there at just about the time that young thing tells me it would."

"What do you think time is?"

"Oh, crap, it's remembering."

"Remembering? Explain."

"Whenever I think of time, I think of something that has either happened or is about to happen. I've been someplace some time ago, and I have someplace to go and I'll be there at some time later. That's what time is, remembering what you did or looking forward to what you will do."

He was making the same argument that Julian Barnes made in his novel *The Sense of an Ending,* that time is just a relationship with memory that "holds us and molds us."[1] Could it be that all long-haul truckers are deep thinkers who multitask watching the road while listening to music and thinking? Are they like the Babylonian shepherds who at one time watched their sheep while thinking about the stars above? They do spend endless hours alone with their thoughts, taking in America as a vast phenomenon of rich observations.

After Rigarro left, I ordered a pie and coffee and stayed on to talk to freight-hauler Bill Moss, a veteran trucker from Georgia who spends ten weeks on the road from Los Angeles to New York and Boston. He rests at home with his family for ten days and then returns to the road for another ten-week, 3,100-mile drive across the country. Bill, who has been trucking since 1985, told me that he never gets bored, unless, of course, he's on that murderously boring stretch through the desert. He drives a monstrously large eighteen-wheeler for Tribe Express Corp., sleeps in his rig, and stops in designated truck stops like FL Roberts Diner.

"I don't think about time at all," he told me. "I enjoy every minute of my time on the big road listening to music, Fox News, and talking on my phone. Time just goes by as fast as the road, when I'm on top. I don't think about the miles and don't think about the time. I know where and when to stop, eat, sleep, shower."

Then I met Phil Major. I climbed into his cab to see the sight from on high.

"Wow! You can see everything from here," I offered.

"Yeah," he admitted, "you get a bit high just from glances of bare legs. That's the kicks we truckers get from being this high."

"Nothing wrong with looking at good-looking things," I said.

I periodically returned to FL Roberts Diner at lunchtimes when it was easy to meet other truckers who were glad to talk. Not knowing any long-haul truckers before my research into time, I had always assumed that truckers would know each other from the old CB radio culture that seems to have died. I introduced myself to a guy sitting at the counter eating a salad and asked the same questions I asked others.

"Road's never boring," he said, not looking at me.

"What about Interstate-40 out of Knoxville?"

"I-40? Hell, you're joking. Knoxville is exciting compared to that same road through the Mojave Desert in southern Nevada. But Christ, even that desert pass is a cool run."

"Cool?"

"Yeah, filled with all sorts of flora. You ever see a Joshua tree? Nothing like it in all America, except in Mojave. Ever see a cougar? I have. The scenery is awesomely beautiful, not boring. There are the wildflowers blooming along the roadside, the rocky terrain, the red sand, and all without traffic. Ask truckers who do the 120-mile stretch through the Mojave Desert across the Colorado River into Arizona and you will get a very different answer. I-40 runs 2,555 miles from California to North Carolina. No. No boring roads in America."

Finn Murphy, the author of the best-selling book *The Long Haul*, had a counterintuitive slant. He told me that there is nothing better than a run where nothing interesting happens. He generally has what he calls an "active inner life" to keep his mind busy while on the road. He keeps to schedule: up by 4:00 a.m., on the road at 5:00 a.m., parked for the night at a truck stop by 4:00 p.m. That's eleven hours, the maximum allowed by federal law, and the maximum a truck's electronic logging would tolerate before signaling an imminent mandatory shutdown. He knows the signs, and knows how much farther it is to the next rest area. He turns off the engine, takes a fifteen-minute power nap, and is renewed and ready to get back on the big slab.

# 17

## 1.5 MILLION YEARS (CIRCADIAN SYNCHRONIZATION)

We are the product of millions of years of living with the sun and moon. So it should not be a surprise to find that artificial tampering with daylight should have consequences for our neurological balance, moods, molecular clockwork, and genetic health. Johnni Hansen, an epidemiologist at the Danish Cancer Society in Copenhagen, studied both daytime and nighttime shift workers and found that women who worked for more than fifteen years reversing day with night slept two hours less and had a significant increase in breast cancer risk over others who worked during normal daylight hours. It seemed to Hansen that by not producing melatonin during the night hours, they were disturbing their circadian rhythms, confusing communications between cells and bodily organs, and therefore compromising their immune systems.[1] He collected information on 1,157 Danish women diagnosed with breast cancer along with data on their work life and risk factors. Several years later, he checked the Danish National Cause of Death Register, discovering that the death rate was 11 percent. By that he concluded that there is a significant tendency of decreasing survival of breast cancer among nightshift workers compared to dayshift workers.

Hansen also studied the effect of artificial light in workplaces on human physiology and behavior. For more than half a century it was known that poor lighting has an adverse effect on mood and that there is substantial evidence that darkness during the day tends to contribute to clinical depression. It seems that disruption of the natural light and darkness rhythms throws off the synchronization of circadian and hormonal rhythms, thus damaging the physiologic and metabolic process and weakening one's ability to sleep and wake with regularity. Sound sleep is known to be beneficial to good health via the immune system. Artificial manipulation of light in a nightshift workplace can curb production of the neurotransmitter serotonin by day, suppress melatonin production and synthesis by night, and muffle biophysical and molecular genetic mechanisms, creating long-term health consequences for workers.[2]

---

Period genes tell us a great deal about how body functions are attuned to the earth's rotations, but it is a long stretch to believe that how they work in the body tells us anything definitive about how humans perceive time. To be sure, our brains are experiencing and processing circadian rhythms, but that does not directly translate back to time perception. Question anyone over the age of fifty about the rate at which the years are passing. The answer is assuredly that years are passing faster and faster with age. There are many good reasons for this. We met one in chapter 13—Paul Janet's psychochronometry model of human temporal impressions of time passing in direct proportion to age. William James was a supporter of that model, so when I was a young man believing that renowned figures knew the absolute truth about what they professed, I too was a supporter. Considering what I know now about the SCN and *per* genes, I find the psychochronometric model conveniently satisfying but far too simple. Here's why.

First, I offer a quotation from the German physician Christoph Wilhelm Hufeland, first physician to the king of Prussia, who back in the late eighteenth century saw the importance of investigating the connection between aging well and an early understanding of practical physiology dictated by circadian rhythms: "That period of twenty-four hours formed by the regular revolution of our earth, in which all its inhabitants partake, is particularly distinguished in the physical economy of man. This regular period is apparent in all diseases; and all the other small periods, so wonderful in our physical history, are by it in reality determined. It is, as it were, the unity of our natural chronology."[3]

The body's circadian system, which we now know involves both the

SCN and *CLOCK* genes in the body, tends to weaken with age. Healthy older humans manage to sleep half as much as young adults.[4] Postmortem studies on Fischer rats show some neuronal deterioration in the SCN, which we know has direct control of circadian rhythm modulations. For several weeks the rats were free to move with implanted sensors telemetering measurements of amplitude reduction and frequency modulation of temperature. Those measurements were seen to be analogous to data observed during the aging process.[5]

It is known that the aging process affects both the SCN and the *CLOCK* genes in the same way that circadian systems are affected by some disease conditions, though healthy aging concession to the circadian sleep-wake cycle at the subcortical structure level is still not understood.[6] In humans, three *per* genes are in the cycle loop that have strong direct effects on the behavior of the whole human organism. Older humans experience decreased duration and quality of sleep. Aging affects neural activity rhythms by disrupting *CLOCK* genes in the circadian pacemaker SCN. And with age, humans experience circadian disturbances through shortened durations of sleep.

All aging things, from living cells to brick houses, need long-term maintenance in their struggles against entropic forces. Living cells approach a limiting time when they can no longer receive and support the transportation of goods and services necessary for living. They follow rates of aging that could be shaped by their local environment, but those rates could also be formed by their genetically programmed depreciation. They gradually lose control of their functions. In a living system that continuously maintains organic energy needed to stabilize the larger organisms that it serves, that rate of aging is called *physiological time,* the kind of time that by some known or unknown mechanism drives living cells to change according to their habitat location and particular time in the universe.

Cells communicate with one another as well as with the outside environment. They sense changes in temperature and can speedily signal that information to other cells by chemical messengers, often changing their own functions. If a cluster of cells is accidentally burned, adjacent cells will detect that their neighbors have been damaged and need repair. Such coordination and response mechanisms allow them to work in teams when accomplishing tasks they could not do on their own, such as hormone release, muscle contraction, and cell migration. Those cells measure time by light intensity and protein buildup and decay. But their response and communication mechanisms begin to wane with age.

It stands to reason that in the past two hundred thousand years of modern human evolution, the entire internal human gene team has been biochemically adapting to the constant and relatively fixed earth orbit and rotation environment. The SCN seems to convey signals to light-sensitive peripheral clocks through a complex feedback system connected by metabolic networks that integrate light-independent systems, including clocks within tissues of the liver, pancreas, skeletal muscle, and intestine.[7] We now have a central pacemaker SCN containing a cell-independent circadian oscillator that synchronizes other brain regions, cells, tissues, and even organs that are complexly tuned to the daily behavioral rhythms of our lives.[8] Loss or damage to that pacemaker results in a desynchronization of circadian clocks at the *per* cell level. Circadian rhythms in *CLOCK* gene- and protein-expressions happen in tissues throughout the body and can adjust to changing environmental signals; moreover, their circadian oscillations continue in cultures, showing a somewhat independence from SCN cells.[9] The clock is within us, masquerading as biological tissue.

Every cell produces and secretes something in a complex circadian rhythm of thousands of genes turning on and off, all working together and communicating with one another to keep the essential functions healthy. The organs know when it's time to eat and sleep, and by their clock they call for organ productions and secretions of whatever, melatonin, insulin, glycogen, to keep all interdependent body functions coordinated. We are filled with clocks. Not only does every cell contain a clock but the liver, stomach, pancreas, heart, and kidney each have an individual clock that must be kept more or less in synch with the others. We adjust our clocks by what we do and by how we respond to the environment; we are more physically active during daylight hours in agreement with the signals sent from our eyes to the SCN to encourage, enforce, and agree with the circadian rhythm encoded in our *per* cells. Hence, physical activity improves sleep cycles; studies show that older adults regularly having measured physical activity improve their sleep quality without the use of sleep medications.[10]

So, could it be that the sense of years passing faster as we age is caused by a complex feedback system from the cells of our bodies to something in our subconscious? Since cells are constantly bombarded by data about the outside world detected via biophysical sensors, it is reasonable to hypothesize that there really is some great metaphorical clock hanging from a picture hook on the hypothalamus guiding the subconscious. Yes, the clock is within us, with an intelligence coming from the outside, seamlessly coordinating the millions of independent things that must work together to protect the whole body. The processes within a cell have influences from beyond cell membranes. So, too, the functions of the brain are influenced by stimuli of a world beyond the boundaries of the skull. Those external stimuli are constantly knocking on the cranium door with useful environmental signals, such as listening to music or going to a baseball game. Every external experience of thought and emotion contributes to an altering of the brain, what the philosophers Andy Clark and David Chalmers call a reliable coupling with the environment. "Once we recognize the crucial role of the environment in constraining the evolution and development of cognition, we see that extended cognition is a core cognitive process, not an add-on extra."[11]

The circadian clock also controls what neurobiologists call oxidative stress, an imbalance of normal metabolic oxygenation activity in cells producing biochemical energy and the release of waste in the routine detoxifying process. That imbalance, which acts in coordination with environmental stresses, generally leads to the production of highly reactive chemicals that can damage normal cells. It has been shown that oxidative stress may lead to chronic and systematic inflammation of cell tissue, instigating biological activity behind cell cancer, dementia, and abnormal aging.

Almost every cell of the body functions in coordination with circadian rhythms, including those in strategic regions of the brain implicated in a few neuropsychiatric diseases such as Parkinson's, Alzheimer's, Huntington's, multiple sclerosis, and ALS (amyotrophic lateral sclerosis). So, even though arrhythmic cell functioning in the brain neither explains nor justifies those diseases, there is some suspicion that maintaining a synchronized rhythm strengthens resilience against them. Satchin Panda, professor in the Regulatory Biology Laboratory at the Salk Institute, tells us in his new book, *The Circadian Code,* that the healthy body keeps all the individual clocks of the body in synchronic alignment. When that alignment is thrown off by social or personal circumstances, the health of the body tends to deteriorate.[12]

The body, through its sense machines, takes in a stridency of noises, impressions of colors, smells and shapes and textures, and in code sends it to the mind, which somehow makes sense of all that noise and swallows the whole mess to magnificently alter the mind by producing something comprehensible.

But the mind has a mind of its own. It knows what to expect from the outside world, so it adjusts its intake to conform to what it has already experienced and to what it knows. It adjusts its perceptions to conform to what it judges as making sense. My personal example comes from a time when I was twelve and lost the sight in my left eye. I was riding my bike home from school when someone threw a stone from across the street. I was told that I would lose depth perception and peripheral perception. For three days I bumped into walls when passing through doorways and reached for things that I thought were nearer than they were. A well-respected physicist told me that depth perception comes from parallax, and of course that parallax relies on two eyes. One other physicist insisted on the absurd belief that seeing in three dimensions required two eyes. Experience tells a different story. It didn't take more than a few days before my brain adjusted itself with external information and a judgment of what makes sense. In just two weeks after the accident I could thread a needle, catch an eighty-mile-per-hour baseball, hit a homerun with a broomstick (on those rare occasions when I could hit a homerun), ride a bike along the top of a six-inch-wide wall, and hit a bull's-eye of a target from a hundred feet away with a BB gun. And with all this, my peripheral vision range was, and still is, about 120 degrees, as wide as anyone's. That's because the eyes see but the brain knows what's real.

The brain has expectations of reality from its experience with the world, and it uses those expectations to judge what to believe. It learns new tasks, and it adjusts to bodily injuries and changes in the environment at remarkable speeds. From experience, habits, and a few hundred thousand years of evolution in a forever-changing environment, it sure knows lots about time and shares it with a few trillion cells in the body. So, we do have a sense of time. The major contributor to our sense of time is surely circadian rhythm; however, significant stimuli come from our communal understanding as well as from our daily habits and routines, and from the bombardment of language involving the most frequently used noun. In the end it is just a noun that begs Augustine's famous question, "What, then, is time?"

## INTERLUDE: TIME IN THE SKY

Captain Richard Sweet pilots for Hawaiian Airlines. He told me that pilots do get sleepy, especially on flights that depart on the back side of their body clocks. "The five-hour flights to the West Coast are so not bad," he said, "but the longer flights from Hawaii to Japan, Korea, Australia, which are ten to eleven hours, are more tiring. Leaving in the afternoon to fly east to New York, the night comes fast and morning comes early." Flying west to Korea is a chase of the sun when he leaves around 12:00 local HST (Hawaii) time. Sweet is tired and ready for bed when he gets there. He feels jet-lagged most of the time. The worst is when he flies east to New York because of the time difference. When he arrives, he sleeps for about six to eight hours, then eats, exercises, and tries to go back to sleep by 10:00 p.m. so that he is ready for a return flight at 7:00 a.m. He has twenty-four hours of rest time. Under his adjusted sleep patterns, getting up at 6:00 a.m. must feel like 12:00 p.m. Hawaii time.

Flying west, the time difference is only four or five hours from Hawaii. So, when he gets to his destination he sleeps well and wakes in the a.m. There is not much jet lag going west. For the sake of his body clock he prefers to fly west. The pilot's slogan is "West is best." His longest flight was ferrying a new Airbus for Hawaiian Airlines, a fun experience. It took sixteen hours from France to Los Angeles. With three other pilots taking turns, he was able to take four-hour breaks. Time passed quickly for the first eight hours, more slowly on the final eight.

The Federal Aviation Administration has strict rules about medication.

Pilots cannot take anything but aspirin or Tylenol. So Sweet resorts to standard jet-lag remedies after getting to his destination, such as drinking lots of water, exercising, and taking a shower. Those remedies help the body to adjust one's body clock to local time.

If everything is working normally, it can be boring on long trips. There is a yearning to get to the destination. Some flights don't go so well. Occasionally, there is a mechanical problem or a passenger with a medical issue. He is then busy, and the time moves fast. Flying at night seems to move more slowly than flying by day. At night, when Sweet feels he has to sleep and can't, time goes slower. He tries to keep busy with recording things like fuel burn, ground speed, wind direction, deviations around storms, and icing levels, watching weather reports on arrival, or doing anything else related to the flight.

"On long flights over eight hours, it seems like time moves slowly. Sometimes it feels as if it stands still when it's smooth and comfortable and everything is going great." He looks at the clock more and anticipates his break. Most pilots find a way to cope with it by reading, talking, doing puzzles, and so on. Some like the long range, some don't. Going home always seems quicker.

# 18

## DISTORTED SENSES AND ILLUSIONS (THE TEMPERATURE OF TIME)

The brain has its storehouse of various clocks linked to cataloging time with memory, planning for the future, the bodily functions of breathing, blood pressure, melatonin secretion, quick motor reactions, and even blinking. Time sorting is essential for the body's survival. A hand touching a hot pot rapidly recoils. Without a fast signal to the brain, the hand might stay long enough on the pot to burn itself into a smelly lump of singed flesh. The mind protects the body it sits in. That's its job.

Those brain-time functions are reasonably locked in a coordinated tempo that coincides with the biochemical necessities of staying alive and healthy. There are, however, other organic factors that contribute to the human sense of time, illusions that are often modified by recent experiences. For example, time seems to slow down when we experience danger or the threat of danger. Awe has the same effect on time perception.[1] Studies on skydiving and bungee jumping suggest that time slows down considerably for people before and during a self-selected dangerous event.

Perhaps that kind of illusion is a remnant of primordial adaptation to the environment of the wild, when the presence of a predator comes within reach of one's personal space. It might be the result of concentrated awareness, when the brain defensibly fixates on a single event to anticipate just how to respond to alternative moves. *If I run, it will come after me. Hmm . . . I might be able to outrun it. Maybe not. If I grunt, it might think I'm stronger than it. If I show my teeth . . .* Each "if" sets the clock back to the beginning of each mental scenario. These ifs have to come in rapid succession, necessitating a modification of temporal awareness. For the entire episode, from the time of becoming aware of the threat to the time of the decision that was made, time slows enormously, giving the impression that it all took a second, when it was very likely ten seconds. Fear builds a temporal illusion that benefits survival. In animals fear creates a hormonal effect in the nervous system that speeds up time while the prey considers the dilemma of whether to fight or take flight. An adrenal hormone torrent of activity creates all sorts of bodily preparations such as enlarging pupils and faster circulation of the blood in preparation for either fight or flight. In that extraordinarily short time interval, the body's normal internal clock is speeded up to accelerate heart and lung action, slow digestion, and release and dedicate all metabolic energy sources for muscular action. This time dilation conforms with, and is supported by, the physiological reaction to threat—bodily changes to fear, pain, hunger, and rage.[2] A great deal of scientific literature demonstrates that angry and fearful cues from the face or the body attract more attention than happy cues.[3]

Hospital patients have distorted senses of time. Time slows in a doctor's waiting room, but it strangely spins in a hospital intensive care unit, where on-and-off sleep tends to muddle reality. One problem is peripheral influences. Human judgment of time depends on all sorts of nontemporal effects of the stimuli, such as what just happened before the judgment of time and the anticipation of what might happen after. Judgment also depends on context, practice, experience, and biasing effects. Recovering from surgery in a hospital bed, whether drugged or drained, severely skews and shrinks time judgment.

In the 1930s the neuroendocrinologist Hudson Hoagland experimented with body temperature and discovered that there is a correlation between body temperature and time perception. He studied electrical impulses that transmit information about sensory processes to the brain from the skin. Since all chemical reactions speed up when heated, Hoagland thought that increases in body temperature might influence time perception for short durations. When his wife, Anna, was ill with a high fever (quite possibly from a case of influenza), he left her for a short time. When he returned, Anna said he had been gone for a long time. Hoagland suspected that her fever sped up her biological clock.[4] He figured that body temperature affects time judgment, just as it does spatial judgment, and that the biological clock has an electrochemical "memory" of the number of milliseconds that pass in any one interval by storing information on the rate at which its chemicals go through modifications.[5]

But Hoagland's conclusions did not consider other external factors, such as the counteracting processes of perspiring and shivering as a result of body heating. We now know that, for example, marijuana decreases body temperature and increases one's sense of time rate. We also know that the prefrontal cortex, which plays a key role in time planning, functions sensitively to brain temperature. There is speculation that neural pathways that regulate our internal clocks are sensitive to internal temperatures. What is not known is whether internal body temperature plays any role in the cognitive processes of attention and memory: two faculties that influence our impressions of time's passing.

By Hoagland's theory, when body temperature increases, all sense of time duration is foreshortened. Some well-documented studies support Hoagland's theory; others do not. In support is the story of the French speleologist (cave explorer and scientist) Michel Siffre, who spent two months alone in underground glaciers of the Alps of southern France in 1962. More than twelve other subterranean expeditions followed, and in 1972, in Del Reo, Texas, he spent six months in a cave with a small group of fellow explorers. In each expedition Siffre and others lived in total isolation without tools to inform them of time's passing. They slept and ate when they felt the need, letting the rhythms of life dictate their impressions of time's passing. Back in 1962, his initial plan was to study the geology of glaciers, but soon he "decided to live like an animal, without a watch, in the dark, without knowing the time."[6] In the 1962 experiment, a team stayed at the entrance of the glacier cave. His feet were always wet, and his body temperature was close to 93°F, but he was able to read, write, and do research on the cave. Siffre signaled to the team whenever he woke, ate, or slept. Every time he signaled, he would count from 1 to 120 at the rate of what he thought would be one digit per second. The record showed that each time he took five minutes to count to 120: his impression of two minutes was really five.

A single bulb lighted Siffre's cave. It was very dark, without any sense of day changing to night or night to day. After two days, his memory lost all reasonable sense of time. He was experiencing one "day" as if it were two. "My sleep was perfect!" Siffre wrote, "My body chose by itself when to sleep and when to eat." He surmised that his sleep and wake cycles were different from what he had experienced on the earth's surface with its daylight and nighttime. His biological clock was adjusting to a different time scheme. In subsequent experiments, other subjects fell naturally into sleep-wake cycles that were more like thirty-six hours of activity and twelve to fourteen hours of sleep. Siffre would sleep for two hours and sometimes for eighteen without being able to tell the difference. Those deep caves provided unspoiled isolation from external time cues that normally connect circadian rhythms to the twenty-four-hour cycle. They seemed to be almost perfect laboratories for controlled experiments to manipulate conditions foreign to the circadian cycle.

In my extensive conversations with astronauts who spent far more time on the International Space Station than Siffre did in caves, the story was different. Perhaps living in extreme darkness for long lengths of time has a marked effect on time sense and circadian rhythms. Living with light is different. None of the astronauts experienced any long-term changes to their circadian rhythms. Though the ISS was alternately moving above the earth's nights and days every ninety minutes, its lights were always on. However, Siffre's cave expeditions were experiments with time sense, whereas astronauts on the ISS were not conducting any tests of time sense. All operations on the station were coordinated in Greenwich Mean Time.

Hoagland's and Siffre's experiments unearthed the question of how different environments change the way humans think of time. Then, in the early 1970s, the German chronobiologist Jürgen Aschoff at the Max Planck Institute for Behavioral Physiology isolated 450 people in a bunker for between three and four weeks with no clock to tell the time of day or night, giving subjects permission to do whatever they liked while their functions were recorded. They lived in a comfortable room with a shower and small kitchen, prepared their own meals, went to bed when tired, and created their own day schedule. Body temperature, activity patterns, bed movements, urine samples, and time estimations were recorded. Communication with the outside world was by sending and receiving letters through a messenger who would deliver supplies through a double door from the outside at random times.

Aschoff found a clear cycle of sleep and wakefulness controlled by body temperature and urine excretion in a rhythm, on average in a period of about twenty-five hours, that was not always in phase with the activity cycle. He himself participated in the study for the first ten days and had this to say in his report:

> After a great curiosity about "true" time during the first 2 days of bunker life, I lost all interest in this matter and felt perfectly comfortable to live "timeless." From the knowledge of animal experiments I was convinced that I had a period shorter than 24 hours; when I was released on day 10, I was therefore highly surprised to be told that my last waking-up time was 3 p.m. In the "mornings," I had difficulty in deciding whether I had slept long enough. On day 8, I got up after only 3 hours of sleep. Shortly after breakfast I wrote in my diary: "Something must be wrong. I feel as if I am on dogwatch." I went to bed again and started the day anew after three more hours of sleep.[7]

So, we wonder about the connection between human metabolic systems and time awareness. Is it possible that metabolic physiological time is connected to universal physical time? Living in caves or living on the International Space Station or jailed in solitary confinement might provide answers.

Humans do not need direct sunlight for needs beyond daily vitamin D requirements; they can live in the dark for long periods. On Baffin Island, above the Arctic Circle, for instance, people live in extended periods of darkness. In late September there is a twilight when the sun disappears from view. By mid-November the region goes into an almost total darkness with the moon giving occasional bits of natural light. In late January twilight emerges, and a few weeks later the sun reappears for an extended length of time. People living in the Arctic experience perpetual darkness for about eleven weeks, with the moon giving a bit of natural light whenever it can.

We are not evolutionarily equipped to cross time zones at speeds out of synch with our circadian rhythms. Fly across a few time zones, east or west, and you are almost certain to experience some degree of *dysrhythmia* (jet-lag fatigue). Dysrhythmia causes both fatigue and hunger. Older people feel it more than younger. You feel out of synch with the time zone, and it does not always depend on how well rested you were on the plane. It seems that the biochemistry in the body—the urine, sodium, glucose, insulin, iron, and potassium levels—continue to be in phase with the time zone of origin. Away from normal environmental timing, your biological rhythms deviate from the normal twenty-four-hour rhythm. Body temperature and organ activity get desynchronized, suggesting that circadian rhythms are endogenous. Those rhythms normally follow the time length of a single day through such environmental synchronizers as light, habit, work, meals, and sleep times, particularly sleep times. A report from the North Atlantic Treaty Organization Advisory Group for Aerospace Research and Development concludes that electromagnetic fields also act as synchronizers, making adjustments more difficult for internal clocks to adapt to sudden shifts of external time sense.[8] A destabilized synchronicity is a direct consequence of confusion between time zones.

Daylight crosses one line every four minutes. With conventional clocks changing every sixty minutes, the day shortens for eastbound flights and lengthens for westbound. Of course, annual variation of daylight and nightlight are convoluted by the tilt of the poles (23.4 degrees) with respect to the ecliptic, the earth's orbit around the sun, giving uneven light cycles to arctic and subarctic regions; however, most transmeridian flights, east or west, do not follow direct meridian-to-meridian crossings. Passengers on those flights will almost always have biological systems out of phase with the environment.

Fortunately, we have zeitgebers, those environmental time cues to endogenous biological rhythms, to help us synchronize our bodies with the environment. We are not plants that have relatively simple biological rhythms synchronized to light, temperature, and other climate-related cycles. Our rhythms are confounded by meals and sleep habits along with work and social routines that keep us synchronized with solar time.[9] And also fortunately, we can help our bodies adjust by making sure that we drink enough nonalcoholic fluids and keep our feet shoeless and elevated on long flights. If not, there is always Ambien or Lunesta for sleep, and Provigil or Nuvigil to wake when we need to. There are many different zeitgebers, the most obvious being light, drugs, temperature, exercise, and eating patterns. They signal to a tiny region in the hypothalamus that concentrations of chemical components in the body should change to accommodate synchronization with the external twenty-four-hour cycle.

## INTERLUDE: MY STRANGE VIEW OF TIME

I have this strange view of the months of the year. I am standing on a large mandala-like circle where the months are marked by twelve paving stones. I am standing on a stone looking at the antipodal stone marking the actual current month. Why such an image? Could some circular calendar of my childhood have made some indelible impression on my date-marking mind? And why am I standing on a stone that marks a time six months before the present month? Shouldn't I already be there at the present month? I shudder to think what a psychoanalyst would make of this. When we get to smaller intervals, such as days of the week, my impressions are far more typical. As with most people who've worked most of their lives in a structured society, Mondays have a far different feeling than Fridays or Sundays. Coming from an Orthodox Jewish background, I still feel Saturdays as significantly different from every other day of the week, even though I left the fold long ago in my teenage years. Hours of the day have changed over my lifetime. In my youth, and even into my early fifties, my typical day would end in the quiet hours of the morning. That all stopped when I stopped full-time teaching. I now rise with daylight, no matter when that daylight comes—around 7:00 a.m. in December, then around 6:00 a.m. in March, then with a quick overnight shift to daylight standard time back to 7:00 a.m. In summer solstice I'm awake at 5:00 a.m. Aside from those waking times, I'm hardly aware of the day's passing time. I look at the clock to keep appointments but am never far from my Google calendar, which flashes, beeps, and pops a window onto my screen an hour before I have to be someplace. This is the result of living in this new electronic age with a dumbed-down feeling of time when we are no longer surrounded by clocks on the streets, in the stores, and in public places. By twilight, I notice that the day has gone by all too fast.

My time zooms. I lop my days into manageable pieces, knowing that almost any job is smaller than the sum of its parts. Overwhelming jobs get done quickly in short, repeated stages. Consider the necessary job of flossing and brushing one's teeth three times a day. In a lifetime the time spent is close to more than a continuous quarter year at eight hours a day! I concede that healthy teeth are worth giving up three months in a lifetime.

Time is what we make of it. I don't speed, because I enjoy life without a broken neck, and so I set my car clock five minutes ahead of the actual time—it always seems to catch me unaware, fools me so that I have an extra five minutes. I do my boring chores in small bites of time. Before long, I feel the accomplishment of doing a huge job with an impression that I didn't spend much time doing it.

# 19

## EXOPLANETS AND BIORHYTHMS (ENVIRONMENTAL SYNCHRONIZERS)

The earth's twenty-four-hour rotation and seasonal cycle are accidents of universal physics, accidents that created the solar system from cosmic dust. With more than a hundred billion galaxies in the observable universe, recent discoveries of earthlike exoplanets orbiting stars—thousands perhaps—suggest that there are many with enough mass to have earthlike atmospheres providing the possibility of life.

Roughly 30 percent of recently discovered red dwarf stars have earth-size, orbiting temperate planets.[1] Proxima Centauri b, which orbits the red dwarf star Proxima Centauri, a mere 4.3 light-years away, was discovered in August 2016, though it has lived for almost five billion years along with its sun. Many of these earthlike exoplanets are locked in orbits, with one side always facing their star, because of their proximity to the star, which causes very strong tides. For organisms living on such a planet, daylight for those living on one side (the side facing its sun) is perpetual; so is nighttime for those living on the side not facing its sun. Moreover, such planets are likely to move in orbits that have eccentricities very different from earth's. Pick any such exoplanet, or perhaps any earthlike extragalactic planet outside our Milky Way galaxy. Its obliquity of ecliptic, its tilt of polar axes, will be different from that of earth, which happens (again, accidentally) to be about 23.4 degrees. That obliquity is responsible for the seasons, if that planet actually has seasons. As that planet orbits its sun, the tilt of its polar axis remains the same against the background of the stars. Therefore, one pole, say its north one, whatever that means, will face away from its sun for half that planet's orbit and face toward its sun for the other half. If the exoplanet's distance to its star changes along the orbit, there may be some environmental biological effect. Activity cycles of almost any red dwarf star might also have an effect. Such a star could brighten or go into ultraviolet and other energetic radiation modes through strong stellar flares that affect the atmosphere of its planet and become detrimental to habitability. All this suggests that any organism living on such a planet would have very different internal clocks from ours, at least as far as sun or sidereal conformity is concerned.

On the other hand, time on such a planet may have nothing to do with its daylight or its nightlight, if it has any. Our earthly organisms have evolved in an environment of the twenty-four-hour day, split between light and dark periods. They have light receptors that can distinguish between light and darkness, and they use that distinction to an advantage. We have vision with a sense of light.

If Proxima Centauri b harbors life (a big if, though possible), a Centauriling's "vision" would most likely be some sense of frequencies at a far end of the human visual spectrum of light, or even radio waves or some superperception that has never been thought of. Centaurilings might sense the position of their planet with respect to their sun in other ways, perhaps by sonar or some sense that we not only don't have but have never thought of. So daily and seasonally life rhythms would be very different from ours for Proxima Centaurilings. A whole year on Proxima happens in just 11.2 earth days.[2] There might not be seasons, and if so, Centaurilings wouldn't need them for food or growing. Life cycles would likely be faster than those on earth. Calendars would simply be markings of time's passing to give a history, an impression of the future, and an order of what-comes-before-what in the playing field of time.

Take this idea one step further by indulging me with the advance of this science fiction fantasy. It is entirely possible that there are intelligent beings on those red dwarf planetary systems. We have been searching for habitable planets in systems having a sun with smaller mass than ours simply because they are easier to find than those in systems with a sun similar in mass to ours. It turns out, however, that we may have been looking at the wrong systems. According to astronomers Shigeru Ida and Feng Tian at Tokyo Institute of Technology, we should be looking in places that are not too hot and not too cold, places with a water-to-land ratio similar to earth's.[3] Forty-two radio telescopes in northern California are paying attention, hoping to detect radio signals from intelligent life around any of the more than ten billion star systems in our Milky Way galaxy.[4] Fifty thousand other star systems are targeted in the planning stages, with data collected by NASA's Transiting Exoplanet Survey Satellite. If intelligent beings on one of those earthlike exoplanets were able to, in some way, make sense of the musical notation of the first five measures of Beethoven's Symphony No. 5 in C minor, Op. 67, would their sense of tempo be consistent with ours?

They would perceive the four-note *short-short-short-long* motif played twice. But would they spot the signature measure of 2⁄4 time, with the calibrated silence and three eighth-notes followed by one half-note, and then the eighth-note of silence between the second and third measures? What they would sense would have nothing to do with their inner clocks calibrated to their planet's sidereal time. And of course, that sidereal time would be very different from ours.

Give those exoplanet beings a viola, or some other musical instrument fitting to their corporeal form and function. In some superintelligent way, teach them our music notation. Have them play the first five measures of Beethoven's Symphony No. 5. Would the beat seem awkward and foreign to us? Would we still recognize the first four pitches as da-da-da-DUM-[eighth beat of silence]-da-da-da-DUM with the beats timed as perfectly on Proxima Centauri b as they would be on earth?

We cannot answer those questions without knowing far more about how those Centaurilings process information in coordination with their own diurnal biorhythms. Very likely, the answers to both questions are negative. We might sense the speed of pitches as a slur with an indistinguishable beat of silence somewhere in the middle. Even though the instructions are directing a mathematical cadence of time, the deciphered effect could miss what we would expect. Moreover, though musical time in the opening measures can be broken into meter, tempo, rhythm, and duration, it is hardly perceived in the same way Beethoven notated it or imagined it. Even earthlings do not have compatible perceptions of those rhythms. They hear it differently in one of two ways, and not as Beethoven himself wrote it. However, suppose we gave Centaurilings a score of repeated notes in regular tempo, perhaps mixing in a few notes of silence. Would they then perceive the intended tempo sensation? I believe the answer is yes, except for a scaling of time.[5]

Daylight hours for those living beings on Proxima Centauri b might not be the critical driving force of their biological rhythms. Perhaps Proxima Centauri b has no atmosphere. An atmosphere is not necessary for life, yet our atmosphere gives us our daylight hours by reflection of the sun's rays on the surface of the earth. Take away the atmosphere and you have eternal night, or at least twilight.

Duration timing for a thinking being on Proxima Centauri b communicating in a conscious language would likely have a duration sense different from a human's; however, its notion of time could be converted to ours by some simple linear transformation. Suppose that Proxima Centauri b is small and spins through its day in twenty earth hours. Suppose that those Proxima Centaurilings had coincidentally taken the Babylonian division of time to be one minute equals sixty parts of an hour. So one minute on Proxima Centauri b is equivalent to five-sixths of a minute (fifty seconds) on earth. We can always do the math in converting Proxima Centauri b time to Greenwich Mean Time on earth and back. Clearly the Proxima Centauriling's biorhythm would be different. Its biorhythm would not be in synch compatible with ours, so its notion of tempo of a musical score might not be the same as ours and might not even be to scale. After all, its notion of time, as in the sense of day length or the feeling of a time span from, say, birthday to birthday would surely be out of phase with ours. Time paces itself according to the body clock that seems to insist that a particular cellular rhythm must be maintained. So an earthling on a visit to Proxima Centauri b would definitely have trouble in the natural sleep cycle of the exoplanet. The body knows time, from its cosmological venue. It ticks away in the pulse and rhythm according to the planetary system of its birth. The body is the elemental clock, the real measure of time. And though we have all sorts of time notions, objective and subjective, scientific and personal, as indicated by a single word, it is the biorhythms and zeitgebers dictated by our own magnificent solar system that give us a measure of time. It's just a measure that neither tells us what time really is nor what it means, for time itself might be solely in the mind.

Time is imagined, a planner of our presence in the world we live in and in the life we lead. It is wrapped in the indefiniteness that always comes with anything without boundaries, a deception that helps us cope with our indefatigable lust for explaining and organizing our world. We might think of it as a limitless line of moments passing off in two directions until it evaporates into emptiness, an infinite or circular maze with no way out, or a densely packed line with points so close together that between any two there is always a third. The mind has trouble representing the finality of the picture, yet something phenomenal does happen in the mind. The more one thinks about the ramifications of time using some sort of visual representation such as the numbers or dials of a clock, the better the mind grasps what those representations actually represent. Intuition about time starts as vague impressions but eventually develops the mind's representative images into a deeper perceptual understanding. I mull over the thought *I have to catch the 9:03 train tomorrow morning.* That contemplation puts a representative number on the path of organized life that I want to live in order to be productive and advance in society. It gives the impression that time is real, having a cause like weight or temperature. The appearance of a clock saying it is 9:03 a.m. connects with an anticipation of the train's arrival, whether it is on time or not. But it is just an appearance of time, not time itself. As Simone Weil wrote in one of her notebooks, "Appearance possesses the fullness of reality, but as appearance only. As anything other than appearance, it constitutes error."[6] Time's only appearance is a ghost of memories and anticipations, a mirage. Without the clock that measures it, tuned to the moves of our little blue planet, it has no existence beyond the biochemical necessity of keeping us alive and its personal effects on memory and destiny.

# EPILOGUE

It seems bottomless. As soon as you think you've uncovered the whole story, you find something that has never been told before. We may never find a definitive answer. After all of our ponderings of the question, however, we may have finally come close; at the very least, we may have figured out how to think about a possible answer that distinguishes contexts surrounding the single word that means so many things. Time, itself, if ever there has been such a thing, could be simply a product of the controlling agents of natural order, the practicalities of the human immersion into an enduring structure of related repetitions and uniformities, the universal mathematics of physics, and the collective accountings of human civilization convolutions.

We rest with the sense that time is just the derivative of a collection of life cycle tasks performed in environments of rhythm and regularity. That's it! It can be so many different things, depending on the measures you explore. We should have different names for it, but the time we most often talk about, that personal time, is just a clock that is part of you and what you do. That would explain why we have no precise sense of time other that what we think we can sense as, say, five minutes when called to dinner. Time might not be real in any other sense than the attempt to organize society by some coordinated clocks that tick away units of instants chosen by human agreement. But what human agreement? We have just come to understand that our cellular clocks—hence, those internal time cycles that dictate half our existences—are somewhat locked and synchronized by the spinning and orbiting of the earth around the sun. Our clever invention, the clock, tells us where to be and what to do, but life is rather dictated by an unyielding harmonization of our planet's orbit and the repeatedly evolving organic zeitgeist at the cellular level. That might be why we are perpetually confused about what time really is.

Physics thinks of time differently. It trusts mathematics, thereby spawning an illusion that time is a thing that cannot be defined in any way other than just being a byproduct of precise mathematical reckonings blithely applied to an indefinite world of human collective behaviors. Because motion is directly connected with space, any change of position is connected to the time of movement or change or gravity, under the assumption that the laws are valid anywhere in the universe. A mathematical model is created to lock time into being a variable that can be manipulated. The *t* that stands for time in a mathematical equation will be the *t* that represents time directly coming from our expediencies of life on a spinning, orbiting earth. However, the physicists boldly extend that representation of time to abstract notions and equations that order the world to predict physical phenomena by laws of physics that must be obeyed.

When we struggle to understand the difference between what physicists call *t* and what we experience as time, we find ourselves siding with the physicists' conception that perhaps time is, indeed, an illusion. It's a logical conclusion, because physics has that way of calling time into mathematical equations, universals, as the tying connection between space and motion, which is explained as being a one-dimensional parameter of spatial movement. It organizes everything that happens in the world by a sequential ordering.

The chronobiologist takes time from organic life and assumes that the measurement of that time is linked to life on earth and its twenty-four-hour cycle of day and night. It means that although organisms have inner times that are very unalike, they are similar in that there is still some universal underlying measure, some root time that impels all measured time. That's a *time* that delineates our collective involvement with life on earth, a *time* that moves with our changing environments, one that brings meaning to our communal existence as a species, a *time* that tells us where we are in life between birth and death.

The philosopher ponders what time *could* be, if there is such a thing.

But we, as a people, live and communicate by appointment and therefore must feel our lives passing from day to day, hour to hour. As it is in Ecclesiastes's long list of times under heaven, there are "true" *times,* when we meet our commitments and play our parts on the stage of society. They are the times we do not completely understand, more universal than simply measuring days, hours, minutes, and seconds. Those times are the byproducts of sidereal and sun times that are merely representative applications of something far more abstract than we can handle. It seems that to know those times at all, one must think of them as only representatives, ambassadors to the real world from a world we probably will never know.

We have all sorts of times: Kantian time, psychological time, physical time, mathematical time, relative time with its paradoxes and dilations, cosmic time with its curiosities, clock time with its simultaneities, and human time with its hastening to the finish. Even fictional characters weigh in. Tony Webster, the narrator of his own past in Julian Barnes's novel *The Sense of an Ending,* gets it when he says, "I know this much: that there is objective time, but also subjective time, the kind you wear on the inside of your wrist, next to where the pulse lies. And this personal time, which is the true time, is measured in your relationship to memory."[1] Paying attention involves the brain, and that attention has everything to do with memory, of how the present becomes the past.

These attempts to answer the big question of time strangely reinforce, complicate, and resist one another. None give a persuasive answer; yet together, they offer a dialectic with no obvious resolution. That big question seems to bring us to a rather fertile endless maze built from a family of answers full of differences and likenesses joined by the *now* in relation to the past and future.

All those times are interrelated, with traces and shades of indeterminable ambiguities that defy strict unity. They battle each other. They buttress each other. They complicate each other, as they all stem from that single word with multiple meanings in discussion with one another. They are a pack of disparate accounts seemingly without any integrated interpretation or resolution and without a unified theory. There should be more than just one word for what we mean by time. However, the times we really care about are simply the ones connected to our human senses; for those, we do have a unified theory built on attention, memory, and perhaps the cellular zeitgeist. We know that kind of time, and we know what it is.

For thousands of years we've been asking the wrong question, searching for some noun that we can envision or conceive, like *weight* or *power*. However, as we have noted in chapter 8, understanding what something *is* is often defined by what it does. So perhaps we should be looking for the verb rather than the noun. A saw is generally a thin, flat blade of tempered steel with a line of triangular teeth mounted to a handle. That is the object. But the object itself is pointless without some indication of what it is used for. Perhaps we should think of time as the thing that moves us, the thing that keeps us going, the thing that makes us feel alive, the recording of life. The clock is the measure of that time, but in the mind that clock is just a mirage that we follow for a sense of achievement, as if it is real. Perhaps that is why time seems to accelerate as we age; we might just be trying to catch up with that mirage before it is too late.

We seem always to think of time by first placing a milestone at the present. Like all animals, we live in the never-ending present. Think of the German psychologist Wolfgang Kohler's experiments of the 1950s that claimed chimpanzees do have some idea of the future, but only measured from the present. We now base our impressions of time only on how civilization dictates its rulings by the clock, as if the clock's precision is something to follow closely in order to survive. Put yourself in the present, and you find yourself almost free from the awareness of time. There is no sense of duration without some relation with our past or with our dreams of the future. Memory is what it takes to bring up the past. Hope and wish fulfillment bring forward the future. Stand in the present, become conscious of time; it will slow down. Pay direct attention to it, meditate on it; it will slow down even more. Time plays tricks with anyone watching it. It is vain and will not let you go. It is happy to hold on to your attention. It grabs on to all those durations stored in the mind that could simply be illusions built from something that does not exist or a mirage that might exist without a possibility that it would ever be discovered, such as the reason for the existence of the universe.

In reality, time is possibly nothing more than an updating of the present, a memory of the past, and an anticipation of the future. That future seems always to be coming forward, like the frames of a film entering the gate of the present before being relegated to the reel of the past. That flow is in our language and hence in the way we think about time.

Or we could think of time the way my granddaughter Lena taught me to think of it. For many Thursdays I would meet her after school for a chat at a local café. On one of those Thursdays, when she was eleven years old and in grade 6 studying the cell, I asked her if she knew about genes and proteins.

"Of course," she said. "They [proteins] are made during the day . . . or night. . . . I forget which, but *they* know what time it is."

"What do you mean?" I asked. "How does a cell know anything about time of day?"

"Your whole body knows," she said assuredly. "Every cell knows the time. Like as if all they have to do is look at a clock on the wall of the brain. . . . That's how we know when it's time to eat. Like, . . . I'm hungry now. Can I have another chocolate croissant?"

We know that the clock is within us and within every living thing. We know that cell clocks in almost all living things are calibrated to the circadian rhythm and are not bothered by relativity. That rhythm is not an illusion. It is real, and unlike Augustine's answer to his own question, we know what it is, most of all that constant circadian rhythm that persistently keeps us tuned and alive. I say, let the physicists and philosophers have their coinciding ideas about what time is so they can explore the mysterious dynamics of subatomic particles, the unfathomable reaches of the universe, and deeper existential questions to understand where we came from and where we are going. For us, though, the answer is much simpler. It is the inner feeling that rides on the rhythm of our cells. We can feel it through our bones with a relaxed acuteness that is good enough for a normal life. When, now and then, we need a sharper sense of time, we look at a clock.

Lena could be right. We needn't look further for the answer to the celebrated question. Look no further. Time is us. We are the clock.

## NOTES

### Preface

1. Lewis Mumford, *Technics and Civilization* (London: Routledge and Kegan Paul, 1934), 17.

### 1. Trickling Waters, Shifting Shadows

Epigraph: Titus Maccius Plautus, *Comedies of Plautus, Translated into Familiar Blank Verse, by the Gentleman Who Translated The Captives,* [translated by Bonnell Thornton] (London: T. Becket and P. A. de Hondt, 1774; facsimile ed., Neuilly-sur-Seine, France: Ulan Press, 2012), 368.

1. George Costard, *The History of Astronomy, with Its Application to Geography, History and Chronology; Occasionally Exemplified by the Globes* (London: James Lister, 1767), 101.

2. Herodotus, *The Histories,* translated by George Rawlinson, vol. 2 (New York: Everyman's Library, 1997), 109.

3. Plautus, *Comedies,* 368–69. Also found in Aulus Gellius, *Attic Nights, Volume 1: Books 1–5,* translated by J. C. Rolfe, Loeb Classical Library 195 (Cambridge, Mass.: Harvard University Press, 1927), 187. There is some speculation that the author might be the second-century BC Roman grammarian Aulus Gellius.

4. Plautus, *Comedies,* 369.

5. Harry Thurston Peck, *Harper's Dictionary of Classical Antiquities* (New York: Harper and Brothers, 1898).

6. Willis Isbister Milham, *Time and Timekeepers, including the History, Construction, Care, and Accuracy of Clocks and Watches* (New York: Macmillan, 1945), 38.

7. See Otto Neugebaur, *A History of Ancient Mathematical Astronomy* (New York: Springer), 609; and Otto Neugebaur, *The Exact Sciences in Antiquity* (New York: Dover, 1969), 81–82.

8. Abd el-Mohsen Bakir, *The Cairo Calendar No. 86637,* Mathạf al-Miṣrī Manu-

script, Papyrus no. 86637, Antiquities Department of Egypt (Cairo: General Organisation for Govt. Printing Offices, 1966).

9. Marshall Clagett, *Ancient Egyptian Science: A Source Book,* vol. 3 (Philadelphia: American Philosophical Society, 1999), 8.

10. Owen Ruffhead, *The Statutes at Large: From the First Year of King Edward the Fourth to the End of the Reign of Queen Elizabeth,* vol. 2 (London: Robert Basket, Henry Woodfall and William Strahan, 1763), 676.

11. Larisa N. Vodolazhskaya, "Reconstruction of Ancient Egyptian Sundials," *Archaeoastronomy and Ancient Technologies* 2, no. 2 (2014): 1–18.

12. The only positive information about the Alexandria clepsydra is a detailed description by the eighteenth-century British politician Charles Hamilton, yet his article contains no citation to check. See Hon. Charles Hamilton, Esq., "A Description of a Clepsydra or Water Clock," *Transactions of the Royal Society of London* 479 (1753): 171.

13. See Vitruvius, *The Ten Books on Architecture,* translated by Morris Hicky Morgan (Cambridge, Mass.: Harvard University Press, 1914), 276.

14. Abraham Rees, *The Cyclopædia: or, Universal Dictionary of Arts, Sciences, and Literature* (London: Longman, Hurst, Rees, Orme and Brown, 1819), 359.

15. Robert Temple, *The Genius of China: 3000 Years of Science, Discovery and Invention* (New York: Simon and Schuster, 2007), 108–9.

16. Ibid.; Joseph Needham, *Science and Civilization in China* (Cambridge: Cambridge University Press, 1965), 446.

17. Derek J. De Solla Price, *On the Origin of Clockwork Perpetual Motion Devices and the Compass* (Australia: Emereo Classics, 2012). This is a Project Gutenberg republication of a 1959 paper that can be found at http://www.gutenberg.org/ebooks/30001.

18. Joseph Mazur, *Zeno's Paradox: Unraveling the Mystery behind the Science of Space and Time* (New York: Plume, 2007), 102.

19. Joshua B. Freeman, *Behemoth: A History of the Factory and the Making of the Modern World* (New York: W. W. Norton, 2018), 1.

20. Ibid., 3.

21. Willis Isbistor Milham, *Time and Timekeepers, including the History, Construction and Accuracy of Clocks and Watches* (New York: Macmillan, 1945).

## Olympic Racer Wins by One Hundredth of a Second

1. NASCAR timing score doesn't measure beyond thousandths of a second, so the official margin of victory is listed as .000 seconds.

## 2. Ringing Bells, Beating Drums

Epigraph: "The 10,000 Year Clock," The Long Now Foundation, www.longnow.org/clock/.

1. Thomas Reid, "Reid's Treatise on Watch and Clock Making," *Watchmaker and Jeweler* 1, no. 1 (1869): 2.

2. "History of the Clock," *Stryker's American Register and Magazine* 4 (July 1850): 351. Note: the present clock has been restored.

3. Henry Sully, *Règle artificielle du temps: Traité de la division naturelle et artificielle du temps*... (1737) (facsimile ed., Whitefish, Mont.: Kessinger, 2010), 272–78.

4. Paul Lacroix, *The Arts in the Middle Ages, and at the Period of the Renaissance* (London: Chapman and Hall, 1875), 175.

5. "Pražský orloj—The Prague Astronomical Clock," http://www.orloj.eu/cs/orloj_historie.htm.

6. Ibid.

7. Philippe de Mézières, *Le Songe du Vieil Pèlerin* (1389), edited by G. W. Coopland, 2 vols. (New York: Cambridge University Press, 1969).

8. Lacroix, *Arts in the Middle Ages,* 174.

9. Dante Alighieri, *The Divine Comedy,* translated by Henry Wadsworth Longfellow (Boston: Ticknor and Fields, 1867), canto 24.

10. I found the evidence for Yi Xing's escapement very tenuous after searching for some corroboration in Joseph Needham's comprehensive book on Chinese science and finding none; Joseph Needham, *Science and Civilization in China* (Cambridge: Cambridge University Press, 1965). So, although I mentioned Yi Xing's escapement, I also post this apprehensive message.

11. Lynn Thorndike, *The "Sphere" of Sacrobosco and Its Commentators* (Chicago: University of Chicago Press, 1949), 1, 230.

12. Needham, *Science and Civilization in China,* 436, 445.

13. "Time's Backward Flight," *New York Times,* November 18, 1883.

14. R. S. Fisher, ed., *Dinsmore's American Railroad and Steam Navigation Guide and Route-Book,* nos. 1–16 (September 1856–December 1857).

15. *Protocols of the Proceedings of the International Conference Held at Washington for the Purpose of Fixing a Prime Meridian and a Universal Day* (Washington, D.C.: Gibson Brothers, 1884), available at http://www.gutenberg.org/files/17759/17759-h/17759-h.htm.

16. Robert Poole, *Time's Alteration: Calendar Reform in Early Modern England* (London: Taylor and Francis, 1998), 1.

17. Calendar Act of 1750, http://www.legislation.gov.uk/apgb/Geo2/24/23.

18. Francis Richard Stephenson and L. V. Morrison, "Long-Term Changes in the Rotation of the Earth: 700 B.C. to A.D. 1980," *Philosophical Transactions of the Royal Society of London A* 313 (1984): 47–70.

19. F. Richard Stephenson, *Historical Eclipses and Earth's Rotation* (Cambridge: Cambridge University Press, 1997), 26.

## 3. Eighth Day of the Week

1. Alberto Castelli, "On Western and Chinese Conception of Time: A Comparative Study," *Philosophical Papers and Reviews* 6, no. 4 (2015): 23–30.

2. I thank my good friend Haiyan Hu for pointing this out to me.

3. Castelli, "On Western and Chinese Conception of Time."

4. Henri Frankfort, *Kingship and the Gods: A Study of Ancient Near Eastern Religion as the Integration of Society and Nature* (Chicago: University of Chicago Press, 1978), 344.

5. G. J. Whitrow, *What Is Time? The Classic Account of the Nature of Time* (Oxford: Oxford University Press, 1972), 5–6.

6. Samuel George Frederick Brandon, *History, Time, and Deity: A Historical and Comparative Study of the Conception of Time in Religious Thought and Practice* (Manchester: Manchester University Press, 1965), 93.

7. Plato, *Collected Dialogues,* edited by Edith Hamilton and Huntington Cairns (Princeton, N.J.: Princeton University Press), 37d, 1168.

8. Alfred R. Wallace, *Man's Place in the Universe* (London: Chapman and Hall, 1904), 310.

9. Mark Twain, *Letters from the Earth: Uncensored Writings,* edited by Bernard DeVoto (New York: Harper Perennial Modern Classics, 2004), 226.

## Part II. Theorists, Thinkers, and Opinions

Epigraph: Moses Maimonides, *The Guide for the Perplexed,* translated by M. Friedlander (London: George Routledge, 1910), 121.

## 4. Zeno's Quiver

1. Shadworth Hollway Hodgson, *Philosophy of Reflection,* vol. 1 (London: Longmans, Green, 1878), 248–54. Available as a 2015 classic reprint from Forgotten Books in London.

2. Joseph Mazur, *Zeno's Paradox: Unraveling the Ancient Mystery behind the Science of Space and Time* (New York: Plume, 2007), 6.

3. Iambichus, *The Life of Pythagoras,* translated by Thomas Taylor (Los Angeles: Theosophical Publishing House, 1918), 63.

4. Plutarch, "On the Sign of Socrates," *Moralia, Volume 7,* translated by Phillip H. De Lacy and Benedict Einarson, Loeb Classical Library 405 (Cambridge, Mass.: Harvard University Press, 1959), 398–99.

5. See Theon of Smyrna, *Mathematics Useful for Understanding Plato,* translated by Robert and Deborah Lawlor (San Diego, Calif.: Wizards Bookshelf, 1979), 1.

6. Aristotle, *The Physics,* translated by Philip H. Wicksteed and Francis M. Cornford, 2 vols. (Cambridge, Mass.: Harvard University Press, 1927, 1934), VIII.7, 2: 369–401.

7. Ibid., II.4, 1: 181.

8. Ibid., IV.11, 1: 383, 385.

9. Lee Smolin, *Time Reborn* (Wilmington, Mass.: Mariner Books, 2014), 83.

10. Lee Smolin, *Three Roads to Quantum Gravity* (New York: Basic Books, 2002), 103.

## Prison for Life without Parole

1. Louis Andriessen, *De Tijd* (1981). The entire work can be heard at https://vimeo.com/77903789.

2. Saint Augustine, *Confessions,* translated by R. S. Pine-Coffin (New York: Penguin, 1961), 262.

3. For more about what it is like to be in the SHU (solitary confinement) without SHU parole, talk to Albert Woodfox, who spent forty-five years in prison, nearly all of it in a fifty-square-foot SHU at the Louisiana State Penitentiary. See Campbell Robertson, "For 45 Years in Prison, Louisiana Man Kept Calm and Held Fast to Hope," *New York Times,* February 20, 2016.

Or learn about Herman Wallace, who spent almost forty-one years in a six-by-nine-foot SHU. Wallace was one of the "Angola 3" placed in solitary confinement for twenty-three hours a day. A 2011 Amnesty International report called Wallace's confinement "a fundamental disregard for his human right." For more about Wallace, see John Schwartz, "Herman Wallace, Freed after 41 Years in Solitary, Dies at 71," *New York Times,* October 4, 2013.

## 5. The Material Universe

Epigraph: Saint Augustine, *Confessions,* translated by R. S. Pine-Coffin (New York: Penguin, 1961), 263.

1. Plato, *The Collected Dialogues of Plato,* translated by Benjamin Jowitt and edited by Edith Hamilton and Huntington Cairns (Princeton, N.J.: Princeton University Press, 1969), 1167.

2. Ibid.

3. William Whewell, "Lyell's Geology Vol. 2," *Quarterly Review* 47 (1832): 126.

4. Charles Lyell, *Principles of Geology,* vol. 1: *An Inquiry How Far the Former Changes of the Earth's Surface Are Referable to Causes Now in Operation* (London: John Murray, 1835), 217.

5. Reijer Hooykaas, *Natural Law and Divine Miracle: The Principle of Uniformity in Geology, Biology, and Theology* (Leiden: E. J. Brill, 1963).

6. John Playfair, *Illustrations of the Huttonian Theory of the Earth* (Edinburgh: Cadell and Davies, 1802), 374.

7. Florian Cajori, "The Age of the Sun and the Earth," *Scientific American,* September 12, 1908.

8. Stephen Jay Gould, *Time's Arrow, Time's Cycle: Myth and Metaphor in the Discovery of Geological Time* (Cambridge, Mass.: Harvard University Press, 1987), 87.

9. John McPhee, *Basin and Range* (New York: Farrar, Straus and Giroux, 1982), 108, 104.

10. Stephen D. Snobelen, "Isaac Newton, Heretic: The Strategies of a Nicodemite," *British Journal for the History of Science* 32, no. 4 (1999): 381–419.

11. Isaac Newton to Robert Bentley, December 10, 1692, 189.R.4.47, ff. 4A–5, Trinity College Library, Cambridge.

12. Sir Isaac Newton, *Newton's System of the World,* translated by Andrew Motte and edited by N. W. Chittenden (New York: Geo. P. Putnam, 1850), 486.

13. Gottfried Wilhelm Leibniz, *Philosophical Essays,* edited and translated by Roger Ariew and Daniel Garber (Indianapolis, Ind.: Hackett, 1989), 329.

14. Augustine, *Confessions,* 263.

15. U.N. Intergovernmental Panel on Climate Change, http://ipcc.ch/report/sr15/.

16. Peter U. Clark et al., "Consequences of Twenty-First-Century Policy for Multi-Millennial Climate and Sea-Level Change," *Nature Climate Change* 6 (2016): 360–69.

17. Shu-zhong Shen et al., "Calibrating the End-Permian Mass Extinction," *Science* 334 (2011): 1367–72.

18. Pallab Ghosh, "Hawking Urges Moon Landing to 'Elevate Humanity,'" *BBC News,* June 20, 2017, https://www.bbc.com/news/science-environment-40345048.

## 6. Gutenberg's Type

1. Edward Gibbon, *The Decline and Fall of the Roman Empire,* vol. 2 (New York: Modern Library, 2003), 530–31.

2. Joseph Mazur, *Zeno's Paradox: Unraveling the Mystery behind the Science of Space and Time* (New York: Plume, 2007), 46.

3. There are several versions of Urban II's speech. They agree in intent but were written several years after the speech was delivered, so there is no way to verify what the pope actually said. Dona C. Munro compares the various texts in "The Speech of Pope Urban II at Clermont, 1095," *American Historical Review* 2 (1906): 231–42.

4. Sister Maria Celeste, *The Private Life of Galileo Compiled Principally from His Correspondence and That of His Eldest Daughter,* edited by Eugo Albéri and Carlo Aruini (Boston: Nichols and Noyes, 1870), 17.

5. Ibid.

6. Robyn Arianrhod, *Einstein's Heroes: Imagining the World through the Language of Mathematics* (New York: Oxford University Press, 2005), 40–41.

7. Sister Maria Celeste, *Private Life of Galileo,* 17.

8. Joseph Mazur, *Zeno's Paradox,* 57–58.

9. Galileo Galilei, *On Motion and on Mechanics,* translated by I. E. Drabkin and Stillman Drake (Madison: University of Wisconsin Press, 1960), 50.

10. Galileo Galilei, *Dialogues concerning Two New Sciences,* translated by Henry Crew and Alfonso De Salvo (New York: Macmillan, 1914), 155.

11. Ibid., 153.

## 7. Enter Newton

1. Sir Isaac Newton, *Newton's Principia: The Mathematical Principles of Natural Philosophy,* translated by Andrew Motte (New York: Daniel Adee, 1846), 78.

2. Katherine Branding, "Time for Empiricist Metaphysics," in *Metaphysics and the Philosophy of Science: New Essays,* edited by Matthew H. Slater and Zanja Yudell (New York: Oxford University Press, 2017), 13–20.

3. Newton, *Principia* 77.

4. Ibid., 81.

5. Ibid.

6. Ernst Mach, *The Science of Mechanics: Account of Its Development,* translated by T. J. McCormack (La Salle, Ill.: Open Court, 1989), 284.

7. Ibid., 32.

8. Henri Poincaré, *The Foundations of Science,* translated by George Bruce Halsted (New York: Science Press, 1913), 227–28.

9. Claude Audoin and Bernard Guinot, *The Measurement of Time: Time, Frequency and the Atomic Clock* (Cambridge: Cambridge University Press, 2001), 11.

## Part III. The Physics

Epigraph: Henri Poincaré, *Science and Hypothesis* (New York: Dover, 1952), 90.

### 8. What Is a Clock?

1. Albert Einstein, "Zur Elektrodynamik bewegter Körper," *Annalen der Physik* 17 (1905): 891.

2. H. A. Lorentz, A. Einstein, H. Minkowski, and H. Weyl, *The Principle of Relativity,* translated by W. Perrett and G. B. Jeffery (London: Methuen, 1923; reprint ed., New York: Dover, 1952), 35–65. Originally published as *Das Relativitätsprinzip,* 4th ed. (Leipzig: Teubner, 1922).

3. Recent studies have assessed the dangers of long-term space flight. See Francine E. Garrett-Bakelman et al., "The NASA Twins Study: A Multidimensional Analysis of a Year-Long Human Spaceflight," *Science* 364 (2019), doi: 10.1126/science.aau8650.

4. See Gerald James Whitrow, *The Natural Philosophy of Time* (Oxford: Clarendon Press, 1980), 265.

5. By frame of reference, we mean a coordinate system with reference points that uniquely locate all points within the system along with a way of measuring distances between points.

6. Indeed, the geometry tells us that using the Pythagorean theorem, we may easily compute the difference between the stationary time interval $\Delta t$ and the moving time interval $\Delta t'$ as $\Delta t' = \Delta t/\sqrt{1 - v^2/c^2}$, where $v$ is the velocity of the moving mirrors to the right. Here is how.

Writing distance as a function of time intervals gives $h = c\Delta t$, $d = v\Delta t'$, and $D = c\Delta t'$.

The Pythagorean theorem applied to the right triangle on page 100 tells us that $h = \sqrt{D^2 - d^2}$.

Therefore, $\Delta tc = \sqrt{(c\Delta t')^2 - (v\Delta t')^2}$.

Solving for $t'$ algebraically, we find that $\Delta t' = \Delta t/\sqrt{1 - v^2/c^2}$. In other words, $\Delta t' > \Delta t$.

7. Lorentz, Einstein, Minkowski, and Weyl, *Principle of Relativity,* 39.

### Time on the International Space Station

1. More precisely, the orbital period of ISS is 92.68 minutes.

### 9. Simultaneous Clocks

1. Henri Poincaré, "La mesure du temps," *Revue de Métaphysique et de Morale* 6 (1898): 1–13.

2. Albert Einstein correspondence with Michele Besso, March 6, 1952.

3. Peter Galison, *Einstein's Clocks, Poincaré's Maps: Empires of Time* (New York: W. W. Norton, 2003), 37, 40.

4. Albert Einstein and Hermann Minkowski, *On the Electrodynamics of Moving Bodies,* translated by M. N. Saha and S. N. Bose (Calcutta: University of Calcutta, 1920), 3.

5. Ibid., 5.

6. Since velocity is distance divided by time, we know that time is distance divided by velocity.

7. Albert Einstein, "On the Electrodynamics of Moving Bodies" (1905), 2, available at https://www.fourmilab.ch/etexts/einstein/specrel/specrel.pdf.

8. Albert Einstein and Michele Besso, *Correspondance avec Michele Besso, 1903–1955* (Paris: Harmann, 1979), 537–38.

9. Some current thinkers suggest that the speed of light at the time of the big bang was far faster than it is now. See Andreas Albrecht and João Magueijo, "Time varying speed of light as a solution to cosmological puzzles," *Physical Review D* 59 (1999): 043516.

10. *The Born-Einstein Letters: Correspondence between Albert Einstein and Max and Hedwig Born from 1916 to 1955 with Commentaries by Max Born,* translated Irene Born (London: Macmillan, 1971), 159.

## 10. Braced Unification

1. Albert Einstein and Hermann Minkowski, *The Principle of Relativity,* translated by Meghnad Saha and S. N. Bose (Calcutta: University of Calcutta Press, 1920), 70.

2. Ibid., 71.

3. Ibid., 71–72.

4. Lorentz's formula for contraction in the direction of motion that contributes to time's dependence on space by the enforced ratio $1{:}\Delta t' = \sqrt{1 - v^2/c^2}$.

5. Eva Brann, *What, Then, Is Time?* (New York: Rowan and Littlefield, 1999), 11.

6. Note here that we are ignoring one of the space dimensions. If we didn't, the light would propagate itself as an expanding sphere. In our case, we chose, for sake of simplicity, to ignore one of the space dimensions. In that case, the light propagates itself as a circle.

7. Collaborative list of authors, "First M87 Event Horizon Telescope Results, I. the Shadow of the Supermassive Black Hole," *Astrophysical Journal Letters,* 875: L1 (April 10, 2019): 1–17.

## 11. Another Midnight in Paris

Epigraph: Ray Bradbury, "A Scent of Sarsaparilla," in *A Medicine for Melancholy and Other Stories* (New York: Bantam, 1960), 61.

1. *Midnight in Paris* (1991), directed by Woody Allen, screenplay available at http://www.pages.drexel.edu/~ina22/splaylib/Screenplay-Midnight_in_Paris.pdf.

2. Ray Bradbury, "A Scent of Sarsaparilla," 61. Thanks to my good friend the poet and philosopher Emily Rolfe Grosholz for pointing me to this story.

3. S. W. Hawking, "Chronology Protection Conjecture," *Physical Review D* 46, no. 2 (1992): 603–11.

4. Lewis Carroll, *Through the Looking-Glass, and What Alice Found There* (London: Macmillan, 1871), chapter 5.

5. Google Maps does give terrain elevation; but, unless you're pedaling a bicycle, you wouldn't give elevation much attention.

6. George Gamow, *One Two Three . . . Infinity: Facts and Speculations of Science* (New York: Viking, 1961), 72.

7. Hawking, "Chronology Protection Conjecture," *Physical Review D* 46, no. 2 (1992): 603–11. Note: Kip Thorne and Sung-Won Kim say otherwise; see Sung-Won Kim and Kip S. Thorne, "Do Vacuum Fluctuations Prevent the Creation of Closed Timelike Curves?," *Physical Review D* 43, no. 12 (1991): 3929–47.

8. Michael S. Morris, Kip S. Thorne, and Ulvi Yurtsever, "Wormholes, Time Machines, and the Weak Energy Condition," *Physical Review Letters* 61, no. 13 (1988): 1446–49.

9. H. G. Wells, *The Time Machine* (1895; reprint, New York: Dover, 1995), 16.

10. J. Richard Gott, *Time Travel in Einstein's Universe: The Physical Possibilities of Travel through Time* (New York: Houghton Mifflin, 2001).

## Part IV. The Cognitive Senses

Epigraph: William James, *Principles of Psychology,* vol. 1 (New York: Macmillan, 1890), 619.

## 12. The Big Question

1. Saint Augustine, *Confessions,* translated by R. S. Pine-Coffin (New York: Penguin, 1961), 14.

2. Immanuel Kant, *Critique of Pure Reason,* translated by F. Max Müller (New York: Macmillan, 1922), 26, 759–60.

3. Geoffrey K. Pullum, *The Great Eskimo Vocabulary Hoax and Other Irreverent Essays on the Study of Language* (Chicago: University of Chicago Press, 1991), 159.

4. Jorge Luis Borges, "A New Refutation of Time" (1946), in *Labyrinths: Selected Stories and Other Writings,* edited by Donald A. Yates and James E. Irby (New York: New Directions, 1962), 205.

5. George Berkeley, *A Treatise concerning the Principles of Human Knowledge* (Mineola, N.Y.: Dover, 2003), 30.

6. William James, *The Principles of Psychology,* vol. 1 (New York: Macmillan, 1890), 244.

7. Ibid., 622.

8. Shadworth Hollway Hodgson, *Philosophy of Reflection,* vol. 1 (London: Longmans, Green, 1878), 248–54.

9. James, *Principles of Psychology,* 605, 606, 607.

10. Ibid., 609.

11. Ibid., 608.

12. Wilhelm Wundt quoted ibid., 626.

13. W. S., "Time," *Kidd's Own Journal; for Inter-Communications on Natural History, Popular Science, and Things in General* 2 (1852): 239.

## 13. Where Did It Go?

Epigraph: Henri Bergson, *Creative Evolution* (New York: Henry Holt, 1911), 16.

1. P. Lecomte du Noüy, *Biological Time* (New York: Macmillan, 1937), 155.

2. William James, *Principles of Psychology,* vol. 1 (New York: Macmillan, 1890), 625.

3. Jean-Claude Dreher et al., "Age-Related Changes in Midbrain Dopaminergic Regulation of the Human Reward System," *Proceedings of the National Academy of Sciences* 105, no. 39 (2008): 15106–11.

4. Warren H. Meck, "Neuropharmacology of Timing and Time Perception," *Cognitive Brain Research* 3, nos. 3–4 (1996): 227–42.

5. P. A. Mangan et al., "Altered Time Perception in Elderly Humans Results from the Slowing of an Internal Clock," *Society for Neuroscience Abstracts* 22, nos. 1–3 (1996): 183.

6. J. E. Roeckelein, "History of Conceptions and Accounts of Time and Early Time Perception Research," in *Psychology of Time,* edited by S. Grondin (Bingley, UK: Emerald Press, 2008), 1–50.

7. Marc Wittmann, *Felt Time: The Psychology of How We Perceive Time,* translated by Erik Butler (Cambridge, Mass.: MIT Press, 2106), 132–34.

8. Du Noüy, *Biological Time,* 164.

## 14. Feeling It

Epigraph: Marcel Proust, *Swann's Way,* translated by C. L. Scott Moncrieff (New York: Henry Holt, 1922), 6.

1. See Dale Guthrie, *The Nature of Paleolithic Art* (Chicago: University of Chicago Press, 2006), vii–x.

2. C. Sinha et al., "When Time Is Not Space: The Social and Linguistic Construction of Time Intervals and Temporal Event Relations in an Amazonian Culture," *Language and Cognition* 3, no. 1 (2011): 137–69.

3. Benjamin Lee Whorf, "An American Indian Model of the Universe," *International Journal of American Linguistics* 16, no. 2 (1950): 67–72.

4. Whorf's work, unfortunately, is mainly based not on fieldwork but rather on a single Hopi living in New York.

5. We should keep in mind that the only professionally ratified Hopi dictionary was compiled at the end of the twentieth century and therefore is tainted by Hopi contacts with English. See Kenneth C. Hill et al., *Hopi Dictionary/Hopìikwa Lavàytutuveni* (Tucson: University of Arizona Press, 1998).

6. Abraham H. Maslow, *Toward a Psychology of Being* (Floyd, Va.: Sublime Books, 2014), 158.

7. Daniel Dennett, *Consciousness Explained* (New York: Little, Brown, 1991), 144.

8. It is interesting to note that the Hopi have no past or future tense; all past or future events for the Hopi are told in a variation of the present.

9. S. G. F. Brandon, "The Deification of Time," in *The Study of Time,* edited by J. T. Fraser, F. C. Haber, and G. H. Müller (Berlin: Springer, 1972), 370–82.

10. R. Núñez and K. Cooperrider, "How We Make Sense of Time," *Scientific American Mind,* 27, no. 6 (2016): 38–43.

11. J. J. Gibson, "Events Are Perceivable but Time Is Not," in *The Study of Time II,* edited by J. T. Fraser and N. Lawrence (New York: Springer, 1975), 295–301.

12. Uma R. Karmarkar and Dean V. Buonomano, "Timing in the Absence of Clocks: Encoding Time in Neural Network States," *Neuron* 53, no. 3 (2007): 427–38.

13. William J. Matthews and Warren H. Meck, "Time Perception: The Bad News and the Good," *WIREs Cognitive Science* 5 (2014): 429–46.

14. Sofia Soares, Bassam V. Atallah, and Joseph J. Paton, "Midbrain Dopamine Neurons Control Judgment of Time," *Science* 354, no. 6317 (2016): 1273–77.

15. Ernst Pöppel, "Lost in Time: A Historical Frame, Elementary Processing Units and the 3-Second Window," *Acta Neurobiologiae Experimentalis* 64 (2004): 295–301.

16. William James, *Principles of Psychology,* vol. 1 (New York: Henry Holt, 2015), 608.

17. Holly Andersen and Rick Grush, "A Brief History of Time Consciousness: Historical Precursors to James and Husserl," *Journal of the History of Philosophy* 47, no. 2 (2009): 277–307; Edmond R. Clay (pseud. Robert Kelly), *The Alternative: A Study in Psychology* (London: Macmillan, 1882), 167–68.

## Undercover at an iPhone Assembly Plant in China

1. Kif Leswing, "Undercover in an iPhone Factory," *Business Insider,* April 11, 2017, https://www.businessinsider.com/qa-with-an-iphone-factory-worker-at-pegatron-changshuo-in-shanghai-2017-4.

## Part V. Living Rhythms

Epigraph: Veronique Greenwood, "The Clocks within Our Walls," *Scientific American,* July 2018, 50–57.

## 15. The Master Pacemaker

Epigraph: Emily Grosholz, "Love's Shadow," from *The Stars of Earth* © Emily Grosholz, 2017, used by permission of Able Muse Press.

1. M. Mila Macchi and Jeffery N. Bruce, "Human Pineal Physiology and Functional Significance of Melatonin," *Frontiers in Neuroendocrinology* 25, nos. 3–4 (2004): 177–95.

2. M. R. Ralph et al., "Transplanted Suprachiasmatic Nucleus Determines Circadian Period," *Science* 247 (1990): 975–78.

3. D. K. Welsh et al., "Individual Neurons Dissociated from Rat Suprachiasmatic Nucleus Express Independently Phased Circadian Firing Rhythms," *Neuron* 14, no. 4 (1995): 697–706.

4. Jonathan Fahey, "How Your Brain Tells Time," *Forbes,* October 15, 2009, https://www.forbes.com/2009/10/14/circadian-rhythm-math-technology-breakthroughs-brain.html#53ea56b23fa7.

5. Mary C. Lobban, "The Entrainment of Circadian Rhythms in Man," *Cold Spring Harbor Symposia on Quantitative Biology* 25 (1960): 325–32.

6. C. A. Czeosler and J. J. Gooley, "Sleep and Circadian Rhythms in Humans," *Cold Spring Harbor Symposia on Quantitative Biology* 72 (2007): 579–97.

## Time on the Trading Floor

1. "The World's Billionaires," *Forbes,* March 5, 2019, https://www.forbes.com/billionaires/#163d2060251c.

## 16. Internal Beat

Epigraph: Lyrics associated with "Theft (or: One Minute Less)," a sixty-second piano composition by Lansing McLoskey (1999). Listen to it at http://www.lansingmcloskey.com/theft.html.

1. Jane A. Brett, "The Breeding Seasons of Slugs in Gardens," *Journal of Zoology* 135, no. 4 (1960): 559–68.

2. Karl von Frisch, *The Dance Language and Orientation of Bees,* translated by Leigh E. Chadwick (Cambridge, Mass.: Harvard University Press, 1967), 253.

3. Martin Lindauer, "Time-Compensated Sun Orientation in Bees," *Cold Spring Harber Symposium of Quantitative Biology* 25 (1960): 371–77.

4. Gerald James Whitrow, *The Natural Philosophy of Time* (Oxford: Clarendon Press, 1980), 135.

5. Theophrastus, *Enquiry into Plants and Minor Works on Odours and Weather Signs,* translated by Sir Arthur Hort (New York: G. P. Putnam's Sons, 1916), 345.

6. William J. Schwartz and Serge Daan, "Origins: A Brief Account of the Ancestry of Circadian Biology," in *Biological Timekeeping: Clocks, Rhythms and Behaviour,* edited by Vinod Kumar (New Delhi: Springer India, 2017), 5.

7. Jean-Jacques de Mairan, "Observation botanique," in *Histoire de l'Académie Royale des Sciences avec les Mémoires de Mathématique et de Physique Tirés des Registres de Cette Académie* (1729), 35.

8. Whitrow, *Natural Philosophy of Time,* 141.

9. Martin C. Moore-Ede, Frank M. Sulzman, and Charles Albert Fuller, *The Clocks That Time Us: Physiology of the Circadian Timing System* (Cambridge, Mass.: Harvard University Press, 1982), 8.

10. Charles Darwin and Francis Darwin, *The Power of Movement in Plants* (New York: D. Appleton, 1881), 402–13.

11. Ronald J. Konopka and Seymour Benzer, "Clock Mutants of *Drosophila melanogaster,*" *Proceedings of the National Academy of Science* 68, no. 9 (1971): 2112–16.

12. Michael Rosbash, "Ronald J. Konopka (1947–2015)," *Cell* 161, no. 2 (2015): 187–88.

13. Paul E. Hardin, Jeffrey C. Hall, and Michael Rosbash, "Circadian Oscillations

in Period Gene mRNA Levels Are Transcriptionally Regulated," *Proceedings of the National Academy of Sciences* 89, no. 24 (1992): 11711–15.

14. Collin S. Pittendrigh, *The Harvey Lectures* (New York: Academic Press, 1961), 95–126.

15. C. B. Saper, G. Cano, and T. E. Scammell, "Homeostatic, Circadian, and Emotional Regulation of Sleep," *Journal of Comparative Neurology* 493, no. 1 (2005): 92–98.

16. Jay C. Dunlap et al., "Light-Induced Resetting of Mammalian Circadian Clock Is Associated with Rapid Induction of the *mPer1* Transcript," *Cell* 91 (1997): 1043–53.

17. M. H. Vitaterna et al., "Mutagenesis and Mapping of a Mouse Gene, Clock, Essential for Circadian Behavior," *Science* 264, no. 5159 (1994): 719–25.

18. *CLOCK* stands for Circadian Locomotor Output Cycles Kaput.

19. For a deeper discussion, see William Bechtel and Adele Abrahamsen, "Decomposing, Recomposing, and Situating Circadian Mechanisms: Three Tasks in Developing Mechanistic Explanations," in *Reduction: Between the Mind and the Brain,* edited by H. Leitgeb and A. Hieke (Frankfurt: Ontos, 2009), 12–177.

20. R. Y. Moore, "Retinohypothalamic Projection in Mammals: A Comparative Study," *Brain Research* 49 (1973): 403–9.

21. F. K. Stephan and I. Zucker, "Circadian Rhythms in Drinking Behavior and Locomotor Activity of Rats Are Eliminated by Hypothalamic Lesions," *Proceedings of the National Academy of Sciences* 69 (1972): 1583–86.

22. S.-I. T. Inouye and H. Kawamura, "Persistence of Circadian Rhythmicity in a Mammalian Hypothalamic 'Island' Containing the Suprachiasmatic Nucleus," *Proceedings of the National Academy of Sciences* 76 (1979): 5962–66.

23. M. R. Ralph et al., "Transplanted Suprachiasmatic Nucleus Determines Circadian Period," *Science* 247, no. 4945 (1990): 975–78.

24. Caroline H. Ko and Joseph S. Takahashi, "Molecular Components of the Mammalian Circadian Clock," *Human Molecular Genetics* 15, suppl. 2 (2006): R271–R277.

25. S. G. Reebs and N. Mrosovsky, "Effects of Wheel Running on the Circadian Rhythms of Syrian Hamsters; Entrainment and Phase Response Curve," *Journal of Biological Rhythms* 4 (1989): 39–48.

26. I want it to be carefully understood that we are not talking about the biorhythm theory, the whimsical hogwash pseudoscience notion that physical, emotional, and mental states are controlled by cycles that depend on one's birthday.

27. Franz Halberg et al., "Transdisciplinary Unifying Implications of Circadian Findings in the 1950s," *Journal of Circadian Rhythms* 1 (2003): 1–61.

## Long-Haul Truckers

1. Julian Barnes, *The Sense of an Ending* (New York: Alfred A. Knopf, 2011), 3.

## 17. 1.5 Million Years

1. Johnni Hansen, "Night Shiftwork and Breast Cancer Survival in Danish Women," *Occupational and Environmental Medicine* 71 (2014): A26.

2. Richard G. Stevens et al., "Meeting Report: The Role of Environmental Lighting and Circadian Disruption in Cancer and Other Diseases," *Environmental Health Perspectives* 115, no. 9 (2007): 1357–62.

3. Christopher William Hufeland, *Art of Prolonging Human Life,* edited by Erasmus Wilson (London: Simpkin and Marshall, 1829), 256.

4. Christina Schmidt, Philippe Peigneux, and Christian Cajochen, "Age-Related Changes in Sleep and Circadian Rhythms: Impact on Cognitive Performance and Underlying Neuroanatomical Networks," *Frontiers in Neurology* 3 (2012), doi: 10.3389/fneur.2012.00118.

5. Germaine Cornelissen and Kuniaki Otsuka, "Chronobiology of Aging: A Mini Review," *Gerontology* 63 (2017): 18–128.

6. Ibid.

7. Jennifer A. Mohawk, Carla B. Green, and Joseph S. Takahashi, "Central and Peripheral Circadian Clocks in Mammals," *Annual Review of Neuroscience* 35 (2012): 445–62.

8. David K. Welsh, Joseph S. Takahashi, and Steve A. Kay, "Suprachiasmatic Nucleus: Cell Autonomy and Network Properties," *Annual Review of Physiology* 72 (2010): 551–77.

9. Mohawk, Green, and Takahashi, "Central and Peripheral Circadian Clocks in Mammals."

10. K. J. Reid et al., "Aerobic Exercise Improves Self-Reported Sleep and Quality of Life in Older Adults with Insomnia," *Sleep Medicine* 11, no. 9 (2010): 934–40.

11. Andy Clark and David Chalmers, "The Extended Mind," *Analysis* 58, no. 1 (1998): 7–19.

12. Satchin Panda, *The Circadian Code: Lose Weight, Supercharge Your Energy, and Transform Your Health from Morning to Midnight* (New York: Rodale Books, 2018), 32.

## 18. Distorted Senses and Illusions

1. Melanie Rudd, Kathleen D. Vohs, and Jennifer Aaker, "Awe Expands People's Perception of Time, Alters Decision Making, and Enhances Well-Being," *Psychological Science* 23, no. 10 (2012): 1130–36.

2. Walter B. Cannon, *Bodily Changes in Pain, Hunger, Fear and Rage: An Account of Recent Researches into the Functions of Emotional Excitement* (Eastford, Conn.: Martino, 2016), 42.

3. Mariska E. Kret et al., "Perception of Face and Body Expressions Using Electromyography, Pupillometry and Gaze Measures," *Frontiers in Psychology* 4 (2013), doi: 10.3389/fpsyg.2013.00028.

4. Hudson Hoagland, *Pacemakers in Relation to Aspects of Behavior* (New York: Macmillan, 1935), 108.

5. Hudson Hoagland, "The Physiological Control of Judgments of Duration: Evidence for a Chemical Clock," *Journal of General Psychology* 9 (1933): 267–87.

6. Joshua Foer and Michel Siffre, "Caveman: An Interview with Michel Siffre," *Cabinet* 30 (Summer 2008), http://www.cabinetmagazine.org/issues/30/foer.php.

7. Jürgen Aschoff, "Circadian Rhythms in Man: A Self-Sustained Oscillator with an Inherent Frequency Underlies Human 24-Hour Periodicity," *Science* 148 (1965): 1427–32.

8. Karl E. Klein and Hans M. Wegmann, "Significance of Circadian Rhythms in Aerospace Operations," North Atlantic Treaty Organization Advisory Group for Aerospace Research and Development, Publication AGARD-AG-247 (AGARD, 1980), 7.

9. Yvan Touitou and Erhard Haus, *Biologic Rhythms in Clinical and Laboratory Medicine* (New York: Springer, 1992), 243.

## 19. Exoplanets and Biorhythms

1. I. Ribas, "A Candidate Super-Earth Planet Orbiting Near the Snow Line of Barnard's Star," *Nature* 563 (2018): 365–68.

2. Rodrigo F. Diaz, "A Key Piece in the Exoplanet Puzzle," *Nature* 563 (2018): 329–30.

3. Feng Tian and Shigeru Ida, "Water Contents of Earth-Mass Planets around M Dwarfs," *Nature Geoscience* 8 (2015): 177–80.

4. Andrew Siemion et al., "Results from the Fly's Eye Fast Radio Transient Search at the Allen Telescope Array," *Bulletin of the American Astronomical Society* 43 (2011): 240.06.

5. Thanks to my conversations with the composer Lansing McLoskey during our joint Bogliasco Fellowship residency, I retracted my belief that exoplanet beings would interpret the long and short beats as just a scaling of our sense of the same. I also learned from McLoskey that nobody hears the opening measures of Beethoven's Symphony No. 5 as Beethoven wrote it and that we perceive those tempos differently.

6. Simone Weil, *Notebooks,* translated by Arthur Wills (London: Routledge and Kegan Paul, 1956), 424.

## Epilogue

1. Julian Barnes, *The Sense of an Ending* (New York: Alfred A. Knopf, 2011), 133.

# ACKNOWLEDGMENTS

My wife, Jennifer Mazur, has always been my inspiration, and most ardent supporter, a person who always gives me honest, constructive criticism. After meticulously reading early manuscript drafts, she wisely suggested editorial changes that clarified all essential concepts. Thank you, Jennifer.

A special thanks goes to Emily Grosholz and Robyn Arianrhod, who also read drafts of the original manuscript, for counseling essential corrections and for excellent editorial advice.

Some compelling questions and potential answers about time came from pondering ancient resources. Most came from hints originating from informed opinions of friends, colleagues, and experts in time sensitivity. Others, very significant ones, came from informative conversations and interviews with long-haul truckers, transcontinental pilots, astronauts at the International Space Station, prisoners experiencing solitary confinement, Olympic racers, clockmakers, and hedge fund traders. That thanks goes to Michael López-Alegría (Commander of NASA's Expedition 14 to the International Space Station), Samantha Cristoforetti (European Space Agency Expedition 33 to the International Space Station), Mike Machado (International Earth Science Constellation Mission Operations Manager), Finn Murphy, Sunita Williams, Jason Hernandez, Nicole Stott, Candito Ortiz, Rodrigo Rigarro, Bill Moss, Phil Major, Richard Sweet, Kif Leswing, Clint Barnum, Dejian Zeng, Damon Rein, Irving Zucker, Jay C. Dunlap, Guillem Anglada Escudé at Queen Mary University of London, Richard Bates (The British Clockmaker), James Levinson, Craig Hammond, Margery Reurink (Vermont Department of Corrections), Brennan Center for Justice, Innocence Project, and Crack Open The Door Project. They are the newly acquired acquaintances who graciously contributed so much of their valuable time in telling their personal experiences with time.

I am exceedingly grateful to the Bogliasco Foundation for a Fellowship and Residency to complete this book at the magnificent Villa dei Pini on the Golfo Paradiso in

Bogliasco, Italy. This exceptionally supportive foundation had generously provided me with ample time, and luxurious comfort and generous service, to bring one to the peak of productivity, all in the company of inspiring fellows, and colleagues, Alf Löhr, Lysley Tenorio, Lansing McLoskey, Bruce Snider, and Sandra Rebok, who contributed both directly and indirectly to the final draft of this book.

I thank the real scholars, philosophers, and scientists who have labored over the question of time to keep the time conversation alive for so many centuries. They were vastly helpful in my research. I also thank Paolo Vian of the Biblioteca Apostolica Vaticana, University of Chicago Digital Preservation Collection, The European Cultural Heritage Online (ECHO), The New York Public Library Digital Gallery, Liberty Fund, PhilSci Archive, Biblioteca Apostolica Vaticana, Archivio di Stato de Firenze, Biblioteca Marucelliana Firenze, Biblioteca Nazionale Centrale Firenze, Università degli Studi di Pavia, *Gallica* (the online rare books library of the Bibliothèque Nationale de France), Scribd.com, Ancientlibrary.com, the Perseus Digital Library, Centro di Ricerca Matematica Ennio De Giorgi, and Biblioteca della Scuola Normale Superiore for allowing me to do research from my home that, a decade ago, would have taken years in rare books rooms of libraries halfway around the world, and to Jonathan Bennett, who maintains a site for the translations of correspondences of mathematicians and philosophers at www.earlymoderntexts.com. And to Gerald James Whitrow, one of the many scholars no longer with us, who have labored over the question of time to uncover what had been intellectually lost for too long.

Thanks to Carol Ann Lobo Johnson for her permission to use her photo of the Prague clock. Thanks to Emily Grosholz and Able Muse Press for their permission to use the poem "Love's Shadow" as a chapter epigraph. And thanks to Tasneem Zehra Husain for permission to use a passage from her forthcoming book *Repaint the Sky.*

Very special appreciations go to my editor, Joseph Calamia, whose insightful suggestions, careful editing, and wonderful restructuring suggestions significantly improved my original manuscript. Along with Joseph Calamia, I thank my dream team of manuscript and page-proof editors at Yale University Press: Laura Jones Dooley, for her expert copyediting and significant role in smoothening out rough paragraphs to better clarify the book's deeper messages; Dan Heaton, senior manuscript editor, for consulting with me and coordinating the last-minute corrections; and Fred Kameny, the proofreader, for diligently reading the final page proofs and catching several errors. And to my agent, Andrew Stewart, whose resolute support carried this project forward from its very brief proposal.

To the Mazur Jefferies family: Catherine, Tom, Sophie, Yelena, and Ned. To the Marshall family: Tamina, Scott, Lena, and Athena. To my brothers, Barry Mazur and Martin Silberberg, my sisters-in-law, Gretchen Mazur, Carole Joffe, and Ruth Melnick, and my brother-in-law, Fred Block; as always, they gave me constant encouragement. To my supportive friends and colleagues who have listened to my many stories; they are constant inspirations for executing good work: Ian Stewart, Tadatoshi Akiba, Jim Tober, Lewis Spratlan, Daniel Epstein, Julian Ferholt, Deborah Ferholt, Marjorie Senechal, and William Cullerne Bown.

# INDEX